SpringerBriefs in Energy

SpringerBriefs in Energy presents concise summaries of cutting-edge research and practical applications in all aspects of Energy. Featuring compact volumes of 50 to 125 pages, the series covers a range of content from professional to academic. Typical topics might include:

- A snapshot of a hot or emerging topic
- A contextual literature review
- A timely report of state-of-the art analytical techniques
- An in-depth case study
- A presentation of core concepts that students must understand in order to make independent contributions.

Briefs allow authors to present their ideas and readers to absorb them with minimal time investment.

Briefs will be published as part of Springer's eBook collection, with millions of users worldwide. In addition, Briefs will be available for individual print and electronic purchase. Briefs are characterized by fast, global electronic dissemination, standard publishing contracts, easy-to-use manuscript preparation and formatting guidelines, and expedited production schedules. We aim for publication 8–12 weeks after acceptance.

Both solicited and unsolicited manuscripts are considered for publication in this series. Briefs can also arise from the scale up of a planned chapter. Instead of simply contributing to an edited volume, the author gets an authored book with the space necessary to provide more data, fundamentals and background on the subject, methodology, future outlook, etc.

SpringerBriefs in Energy contains a distinct subseries focusing on Energy Analysis and edited by Charles Hall, State University of New York. Books for this subseries will emphasize quantitative accounting of energy use and availability, including the potential and limitations of new technologies in terms of energy returned on energy invested. The second distinct subseries connected to SpringerBriefs in Energy, entitled Computational Modeling of Energy Systems, is edited by Thomas Nagel, and Haibing Shao, Helmholtz Centre for Environmental Research - UFZ, Leipzig, Germany. This sub-series publishes titles focusing on the role that computer-aided engineering (CAE) plays in advancing various engineering sectors, particularly in the context of transforming energy systems towards renewable sources, decentralized landscapes, and smart grids.

All Springer brief titles should undergo standard single-blind peer-review to ensure high scientific quality by at least two experts in the field.

Junwei Shen · Shaowu Ma · Yuning Zhang ·
Jian Chang

High-Speed Photography in Fluid Mechanics

Junwei Shen
Key Laboratory of Power Station Energy
Transfer Conversion and System, Ministry
of Education
North China Electric Power University
Beijing, China

Shaowu Ma
Key Laboratory of Power Station Energy
Transfer Conversion and System, Ministry
of Education
North China Electric Power University
Beijing, China

Yuning Zhang
Key Laboratory of Power Station Energy
Transfer Conversion and System, Ministry
of Education
North China Electric Power University
Beijing, China

Jian Chang
Key Laboratory of Power Station Energy
Transfer Conversion and System, Ministry
of Education
North China Electric Power University
Beijing, China

ISSN 2191-5520 ISSN 2191-5539 (electronic)
SpringerBriefs in Energy
ISBN 978-3-031-82753-2 ISBN 978-3-031-82754-9 (eBook)
https://doi.org/10.1007/978-3-031-82754-9

This book was financially supported by the National Natural Science Foundation of China (Project No.: 51976056).

© The Editor(s) (if applicable) and The Author(s), under exclusive license to Springer Nature Switzerland AG 2025

This work is subject to copyright. All rights are solely and exclusively licensed by the Publisher, whether the whole or part of the material is concerned, specifically the rights of translation, reprinting, reuse of illustrations, recitation, broadcasting, reproduction on microfilms or in any other physical way, and transmission or information storage and retrieval, electronic adaptation, computer software, or by similar or dissimilar methodology now known or hereafter developed.
The use of general descriptive names, registered names, trademarks, service marks, etc. in this publication does not imply, even in the absence of a specific statement, that such names are exempt from the relevant protective laws and regulations and therefore free for general use.
The publisher, the authors and the editors are safe to assume that the advice and information in this book are believed to be true and accurate at the date of publication. Neither the publisher nor the authors or the editors give a warranty, expressed or implied, with respect to the material contained herein or for any errors or omissions that may have been made. The publisher remains neutral with regard to jurisdictional claims in published maps and institutional affiliations.

This Springer imprint is published by the registered company Springer Nature Switzerland AG
The registered company address is: Gewerbestrasse 11, 6330 Cham, Switzerland

If disposing of this product, please recycle the paper.

Preface

High-speed photography technology shows significant scientific importance in fluid mechanics. It enables precise capture and visual analysis of the fluid motion. This book provides a detailed introduction to the development history and application scenarios of the high-speed photography technology in fluid mechanics.

Based on high-speed imaging technology, experimental observations and underlying mechanisms have been conducted on different sub-division of fluid mechanics, including the bubble dynamics, drop dynamics, and wake dynamics. For the bubble dynamics, firstly, the morphological evolution, collapse jet, shock wave and other characteristics of the bubble collapsing near boundaries are analyzed. Secondly, the effects of the bubble oscillation on the particle motion are analyzed, and the acceleration mechanisms of the particle are explored. Finally, the flow characteristics and structures under various cavitating flow patterns are investigated.

For the drop dynamics, firstly, the morphological variations are discussed when the drops impact onto different surfaces, which include thin liquid layers, wet surfaces, and dry surfaces. Secondly, the coalescence behaviors of the drop-flat interface and drop-drop are explored. Finally, various types of drop fragmentation mechanisms are analyzed, including drop splashing, pinch-off, drop falling fragmentation, and metal drop fragmentation.

For the wake dynamics, the development and evolution of the wake during the translational and rotational motions of the blunt bodies are analyzed, including cylinders, spheres, and prisms.

Beijing, China
November 2024

Junwei Shen
Shaowu Ma
Jian Chang
Yuning Zhang

Contents

1 **Introduction** 1
   1.1 Research Background 1
   1.2 Research Status 2
   1.3 Description of the Book 4
   References 5

2 **High-Speed Photography Technology** 7
   2.1 Introduction 7
   2.2 High-Speed Imaging 7
      2.2.1 Development History 7
      2.2.2 Application Scenarios 8
   2.3 Typical High-Speed Photography Experimental System 9
      2.3.1 Visualization Platform for Bubble Oscillation 9
      2.3.2 Schlieren Method Imaging Platform 10
      2.3.3 Water/Wind Tunnel Platform 11
   2.4 Future Directions 12
   References 12

3 **Visualization Research on Bubble Dynamics** 15
   3.1 Introduction 15
   3.2 Spherical Bubble Near Various Boundaries 15
      3.2.1 Bubble Deformation 16
      3.2.2 Collapse Jet 18
      3.2.3 Shock Wave 20
   3.3 Bubble Within Narrow Gaps 22
      3.3.1 Flat Boundary 22
      3.3.2 Curved Boundary 23
   3.4 Particle-Bubble Interaction 25
      3.4.1 Single Particle 26
      3.4.2 Multi-particle 27
   3.5 Cavitation Flow 29
      3.5.1 Sheet Cavitation 29

	3.5.2 Cloud Cavitation	30
	3.5.3 Super Cavitation	33
	References	35

4 Visualization Research on Drop Dynamics 37
  4.1 Introduction 37
  4.2 Drop Impact and Bouncing 37
     4.2.1 Thin Liquid Layer 37
     4.2.2 Wetted Surface 40
     4.2.3 Dry Surface 41
  4.3 Drop Coalescence 44
     4.3.1 Drop-Flat Interface 44
     4.3.2 Drop-Drop 46
  4.4 Drop Fragmentation 48
     4.4.1 Drop Splashing 48
     4.4.2 Drop Pinch-Off 51
     4.4.3 Drop Falling Fragmentation 52
     4.4.4 Metal Melt Drop Fragmentation 53
  References 55

5 Visualization Research on Blunt Body Wake Dynamics 57
  5.1 Introduction 57
  5.2 Cylinder 57
     5.2.1 Translation 57
     5.2.2 Rotation 59
  5.3 Sphere 60
     5.3.1 Rolling 60
  5.4 Prism 64
     5.4.1 Translation 64
  References 69

6 Conclusion 71

Index 73

About the Authors

Junwei Shen is a doctoral candidate at North China Electric Power University. His research interest focuses on the cavitation and bubble dynamics within confined space. He has published 13 journal papers based on high-speed photography for investigating the bubble jet and interface motion.

Shaowu Ma is a doctoral candidate at North China Electric Power University. He mainly focuses on the research of the cavitation bubble collapse dynamics near boundaries. Based on the high-speed photography technology, he has achieved multi-view visualization on the bubble deformation and movement characteristics, with a maximum shooting speed of 1 million fps.

Yuning Zhang is a professor at North China Electric Power University. He primarily focuses on the research in cavitation and bubble dynamics. He has published 3 monographs in Springer Press and over 90 papers in journals such as Nature Communications, Physics of Fluids, and Energy. He was selected as one of the "top 2% of scientists in the world", released by Stanford University. In addition, he is an Associate Editor of IET Renewable Power Generation and an Editorial Board Member of 6 international/national journals including Journal of Hydrodynamics.

Jian Chang is a professor at North China Electric Power University. He is mainly engaged in the research on multiphase flow and reaction engineering. He has published over 30 papers in journals such as AIChE Journal, Chemical Engineering Science, Industrial & Engineering Chemistry Research, and Powder Technology. He serves as section chairman of the annual meeting 2023 of Chemical Industry and Engineering Society of China and the 10th China Fluidization Conference. He received an award from the Ministry of Education of the People's Republic of China.

Chapter 1
Introduction

1.1 Research Background

In fluid mechanics, high-speed photography is a technique for capturing the instantaneous interface dynamics [1–3]. By continuously capturing images within a short time, it enables the recording of momentary specifics during the motion of the fluid interface. Based on this technology, one can obtain a complete physical process of the rapid motion.

High-speed photography is widely applied in numerous fields. For instance, in industrial production, it is employed for equipment condition monitoring [4]; in the military sector, it records significant processes such as missile launches and explosion experiments [5]; and in sports, it captures athletes' dynamics and skills during the training [6]. Especially in scientific research, high-speed photography aids scientists in uncovering the mechanisms of rapid physical phenomena, thereby advancing the scientific research [7–9].

In fluid mechanics, high-speed photography has applied for exploring the internal cavitation erosion of the fluid machineries [10–12], atomization phenomena [13–15], and vortex shedding phenomena [16–18]. Specifically, exploring the mechanisms of the cavitation erosion can improve the performance and prolong the service life of the hydraulic machinery, thereby ensuring its safe and stable operation and optimizing its design and material selection [19]. Research on the atomization phenomena can advance atomization technology, which is significant for improving fuel combustion efficiency [20]. Atomization technology also plays a vital role in pesticide spraying and industrial coating processes. In the engineering field, vortex shedding phenomena exert a considerable influence on the stability of the equipment such as the aircraft and water turbines [21]. Figure 1.1 illustrates specific applications of the high-speed photography in fluid mechanics. In the figure, various high-speed phenomena are explored: drop fragmentation after impacting a rigid wall (i.e., subfigure (a)); cavitation bubble collapse and liquid jets near a cylinder (i.e., subfigure (b)); vortex

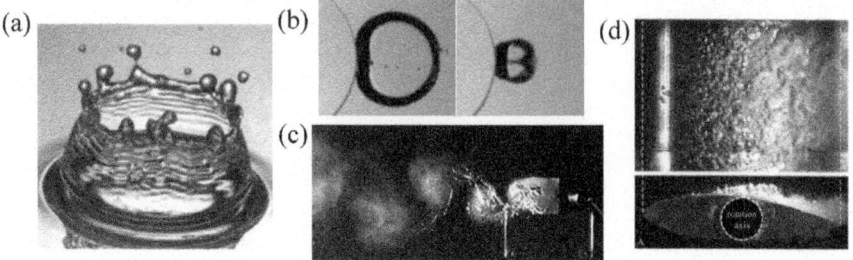

Fig. 1.1 Applications of the high-speed photography in fluid mechanics. **a** Drop impact, reprinted with the permission from Ref. [22] Copyright (2021) (ELSEVIER). **b** Bubble collapse. **c** Wake of a blunt body, reprinted with the permission from Ref. [23] Open access under a CC BY 4.0 license, https://creativecommons.org/licenses/by/4.0/. **d** Cavitation flow, reprinted with the permission from Ref. [24] Copyright (2022) (ELSEVIER)

shedding from the trailing edge of a bluff body (i.e., subfigure (c)); and cavitation erosion on the hydrofoil surface (i.e., subfigure (d)).

1.2 Research Status

The application of the high-speed photography in fluid mechanics mainly includes three objects: bubbles, drops, and wakes. Table 1.1 shows the current research status in fluid mechanics employing the high-speed photography.

For the visualization research on the bubble dynamics, the primary phenomena include: bubble deformation, collapse jet, shock wave, particle motion, and cavitation flow. Bubble deformation and collapse jet are closely related, reflecting the intensity of the bubble collapse, with notable variations observed in proximity to different boundaries. Zhang et al. [25] observed heart-shaped bubbles collapsing near a flat

Table 1.1 Current research status in fluid mechanics employing the high-speed photography

Object	Phenomenon	Refs
Bubble	Bubble deformation	[25–33]
	Collapse jet	
	Shock wave	
	Particle motion	
	Cavitation flow	
Drop	Impact	[22, 34–39]
	Coalescence	
	Fragmentation	
Wake	Translational wake	[40–44]
	Rotational wake	

1.2 Research Status

wall, and discovered typical phenomena of the jet and counter-jet. Specifically, during the rebound period, the collapse jet generated towards the wall propels the bubble towards it, which is then followed by a counter-jet that pushes the bubble away from the wall. Brujan et al. [26] conducted more detailed high-speed photographic research on the bubble behaviors near a right-angle wall. Results showed that at a large bubble-wall distance, an asymmetric annular bubble is formed after the jet penetrates the bubble, which is perpendicular to the jet direction. And, it collapses radially and moves towards the vertex of the wall. At a small bubble-wall distance, the bubble collapses in a crescent shape in the second oscillation period. Tomita et al. [27] explored the dynamics of the bubble deformation and collapse jet near convex and concave walls. The results showed that the wall curvature significantly affects the bubble deformation. The collapse jet velocity near a convex wall is significantly higher than that near a flat wall.

Shock waves are one of the primary mechanisms contributing to cavitation erosion. Lindau and Lauterborn [28] investigated the bubble oscillation phenomena near a flat wall and demonstrated the shock wave evolution through the schlieren techniques. The jet impacts the lower bubble interface and generates a circular shock wave around the bubble. And, it becomes more pronounced towards the end of the collapse. Požar et al. [29] conducted the research on shock waves in the proximity of the concave wall. In contrast to flat and convex walls, the shock waves near a concave wall undergo refocusing, resulting in secondary cavitation. For the particle motion, Li et al. [30] systematically investigated the bubble-particle interactions through the electrical spark experiments. The research indicated that the increasing particle-bubble radius ratio leads to a weakening of the acceleration effect of the particle. And, an increase in the ratio of particle to liquid density resulted in a decrease in particle velocity. Additionally, they also observed a secondary acceleration effect of the particle. For the cavitating flows, Skripkin et al. [31] explored the development characteristics of the two-dimensional cavitation near an NACA (National Advisory Committee for Aeronautics) 0012 hydrofoil. Based on different liquid acceleration times, they identified two distinct cavitation flow states around the hydrofoil: unsteady and quasi-steady.

For the visual research on the drop dynamics, the primary phenomena include drop impact, coalescence, and fragmentation. Regarding the drop impact, Okawa et al. [22] investigated the splash phenomenon of the drops impacting on the thin liquid layer, and discovered two typical splashes: early and late splashes. The research indicated that the hydraulic properties of the thin liquid layer affect the early splash, regardless of the liquid layer thickness. Late splash is affected by both the hydraulic properties and liquid layer thickness. Zhuo et al. [34] investigated the drop impact behaviors on dry wooden surfaces and explored the effects of the surface roughness. They found that when a drop impacts on a dry surface, it forms a circular liquid film. The increasing surface roughness reduces the liquid film fluidity, manifesting as decreased spreading and more vigorous drop splashing. Additionally, the physico-chemical properties of the wooden surface also affect the drop impact behaviors. As for the drop coalescence, Chashechkin et al. [35] investigated the drop coalescence with a free surface. They applied colored drops to coalesce with a free liquid surface at different contact velocities. The results showed that at low contact velocities, the drop immerses in a dense lenticular shape and subsequently develops into sinking vortex rings. At high contact velocities, fast jets are formed and spread outwards, forming a network structure on the liquid surface. Regarding the drop fragmentation, Avila and Ohl [36] investigated the influence mechanisms of the drop fragmentation, and observed three typical fragmentation phenomena: rapid atomization, sheet formation, and coarse fragmentation. At a large bubble-drop radius ratio, the bubble caused rapid rupture of the drop, leading to the atomization. At a medium bubble-drop radius ratio, the drop transformed into the liquid films, which contracted or broke up in sheets. At a small bubble-drop radius ratio, numerous high-speed jets were formed on the drop surface, resulting in coarse breakup.

For the visual research on the wake dynamics, it primarily includes the wakes of the blunt body translation and rotation. For the translational motion of the bluff bodies, Veldhuis et al. [40] investigated the wake behaviors of a sphere undergoing upward or downward translational motion, and discovered typical structures such as the double-threaded structure and hairpin-like vortices. Shen et al. [41] explored the wake dynamics of the cylinder with a flexible film undergoing relative translational motion in a flow field. They found that the flexible film significantly affects the wake structure, with a greater effect as its length increases. When the ratio of the flexible film length to cylinder diameter was 5, it could effectively suppress the cylinder wake. Thiria et al. [42] investigated the wake dynamics of the cylinder undergoing rotational oscillatory motion.

1.3 Description of the Book

The main chapters of the book are organized as follows: This chapter presents the application scenarios of the high-speed photography in fluid mechanics, and reviews the research on the high-speed photography in bubble, drop, and wake dynamics. Chapter 2 introduces the development history of the high-speed photography technology, as well as its application scenarios in experimental fluid mechanics and the future prospects. Chapter 3 introduces the dynamic behaviors of the bubble in various scenarios, including spherical bubble, bubble within narrow gaps, bubble-particle interaction, and cavitation flow. Chapter 4 introduces the dynamic characteristics of the drop impact and bounce at different interfaces, and explores the coalescence of the drop-flat interface and the drop-drop. In addition, various types of the drop fragmentation mechanisms are analyzed, including the drop splashing, pinch off, drop falling fragmentation, and metal drop fragmentation. Chapter 5 introduces the high-speed photographic visualization research on the wake dynamics, including the translational and rotational wake of the blunt bodies, such as cylinders, spheres, and prisms. Chapter 6 sums up the main conclusions of the book.

References

1. Versluis M (2013) High-speed imaging in fluids. Exp Fluids 54:1–35
2. Fuller PWW (2009) An introduction to high speed photography and photonics. Imag Sci J 57(6):293–302
3. Field JE (1983) High-speed photography. Contemp Phys 24(5):439–459
4. Lin X, Zhu K, Fuh JYH et al (2022) Metal-based additive manufacturing condition monitoring methods: from measurement to control. ISA Trans 120:147–166
5. Mizushima Y (1979) High-speed photography of explosion in Japan. In: 13th international congress on high speed photography and photonics. SPIE, vol 189, pp 109–114
6. Wilson BD (2008) Development in video technology for coaching. Sports Technology 1(1):34–40
7. Yamada H, Sato T, Fujiwara T (1990) High-speed photography of prebreakdown phenomena in dielectric liquids under highly non-uniform field conditions. J Phys D Appl Phys 23(12):1715
8. Bourne NK, Obara T, Field JE (1997) High-speed photography and stress gauge studies of jet impact upon surfaces. Philos Trans Royal Soc London. Series A: Math, Phys Eng Sci 355(1724):607–623
9. Dewey JM, Kleine H (2005) High-speed photography of microscale blast wave phenomena. In: 26th international congress on high-speed photography and photonics. SPIE, vol 5580, pp 106–114
10. Dular M, Petkovšek M (2015) On the mechanisms of cavitation erosion–Coupling high speed videos to damage patterns. Exp Thermal Fluid Sci 68:359–370
11. Hutli E, Nedeljkovic MS, Radovic NA et al (2016) The relation between the high speed submerged cavitating jet behaviour and the cavitation erosion process. Int J Multiph Flow 83:27–38
12. Lv D, Lian Z, Zhang T (2018) Study of cavitation and cavitation erosion quantitative method based on image processing technique. Adv Civ Eng 2018(1):5317578
13. Koračin N, Zupančič M, Vrečer F et al (2022) Characterization of the spray droplets and spray pattern by means of innovative optical microscopy measurement method with the high-speed camera. Int J Pharm 629:122412
14. Liu AB, Reitz RD (1993) Mechanisms of air-assisted liquid atomization. Atomiz Sprays 3(1)
15. Charalampous G, Hadjiyiannis C, Hardalupas Y et al (2010) Measurement of continuous liquid jet length in atomizers with optical connectivity, electrical conductivity and high-speed photography techniques. In: Proceedings of 23rd annual conference on liquid atomization and spray systems, ILASS—Europe
16. Peng C, Tian S, Li G (2021) Determination of the shedding frequency of cavitation cloud in a submerged cavitation jet based on high-speed photography images. J Hydrodyn 33(1):127–139
17. Ausoni P (2009) Turbulent vortex shedding from a blunt trailing edge hydrofoil. Epfl
18. Dye RCF (1978) Photographic evidence of the mechanisms of vortex-excited vibration. J Photogr Sci 26(5):203–208
19. Yılmaz Ö, Aksoy M, Kesilmiş Z (2022) Investigation of the relationship between vibration signals due to oil impurity and cavitation bubbles in hydraulic pumps. Electronics 11(10):1549
20. Faik AMD, Zhang Y (2018) Multicomponent fuel droplet combustion investigation using magnified high speed backlighting and shadowgraph imaging. Fuel 221:89–109
21. Rashidi S, Hayatdavoodi M, Esfahani JA (2016) Vortex shedding suppression and wake control: a review. Ocean Eng 126:57–80
22. Okawa T, Kubo K, Kawai K et al (2021) Experiments on splashing thresholds during single-drop impact onto a quiescent liquid film. Exp Thermal Fluid Sci 121
23. Wu J, Deijlen L, Bhatt A et al (2021) Cavitation dynamics and vortex shedding in the wake of a bluff body. J Fluid Mech 917
24. Nichik MY, Timoshevskiy MV, Pervunin KS (2022) Effect of an end-clearance width on the gap cavitation structure: Experiments on a wall-bounded axis-equipped hydrofoil. Ocean Eng 254

25. Zhang J, Du Y, Liu J et al (2022) Experimental and numerical investigations of the collapse of a laser-induced cavitation bubble near a solid wall. J Hydrodyn 34(2):189–199
26. Brujan EA, Noda T, Ishigami A et al (2018) Dynamics of laser-induced cavitation bubbles near two perpendicular rigid walls. J Fluid Mech 841:28–49
27. Tomita Y, Robinson PB, Tong RP et al (2002) Growth and collapse of cavitation bubbles near a curved rigid boundary. J Fluid Mech 466:259–283
28. Lindau O, Lauterborn W (2003) Cinematographic observation of the collapse and rebound of a laser-produced cavitation bubble near a wall. J Fluid Mech 479:327–348
29. Požar T, Agrež V (2021) Laser-induced cavitation bubbles and shock waves in water near a concave surface. Ultrason Sonochem 73:105456
30. Li S, Zhang A, Wang S et al (2018) Transient interaction between a particle and an attached bubble with an application to cavitation in silt-laden flow. Phys Fluids 30(8)
31. Skripkin SG, Tsoy MA, Kravtsova AY (2022) Experimental study of cavitating flow around a NACA 0012 hydrofoil in a slit channel. Sci Rep 12(1):11182
32. Soh WK, Willis B (2003) A flow visualization study on the movements of solid particles propelled by a collapsing cavitation bubble. Exp Thermal Fluid Sci 27(5):537–544
33. Li S, Zhang AM, Han R et al (2017) Experimental and numerical study on bubble-sphere interaction near a rigid wall. Phys Fluids 29(9)
34. Zhuo YY, Hussain S, Lin SY (2021) Effect of surface roughness on the collision dynamics of water drops on wood. Colloids Surf, A 612:125989
35. Chashechkin YD, Ilinykh AY (2023) Intrusive and impact modes of a falling drop coalescence with a target fluid at rest. Axioms 12(4):374
36. Avila SRG, Ohl CD (2016) Fragmentation of acoustically levitating droplets by laser-induced cavitation bubbles. J Fluid Mech 805:551–576
37. Sa R, Takahashi M, Moriyama K (2011) Study on fragmentation behavior of liquid lead alloy droplet in water. Prog Nucl Energy 53(7):895–901
38. Thoroddsen ST, Etoh TG, Takehara K (2008) High-speed imaging of drops and bubbles. Annu Rev Fluid Mech 40(1):257–285
39. Lardier N, Roudier P, Clothier B et al (2019) High-speed photography of water drop impacts on sand and soil. Eur J Soil Sci 70(2):245–256
40. Veldhuis C, Biesheuvel A, Van Wijngaarden L et al (2004) Motion and wake structure of spherical particles. Nonlinearity 18(1):C1
41. Shen P, Lin L, Wei Y et al (2019) Vortex shedding characteristics around a circular cylinder with flexible film. Eur J Mech-B/Fluids 77:201–210
42. Thiria B, Goujon-Durand S, Wesfreid JE (2006) The wake of a cylinder performing rotary oscillations. J Fluid Mech 560:123–147
43. Chandel A, Das SP (2021) Effect of wall proximity on the wake of a rotating and translating sphere. Acta Mech 232:4833–4846
44. Chandel A, Das SP (2021) Wake of transversely rotating and translating sphere in quiescent water at low Reynolds number. Acta Mech 232:949–966

Chapter 2
High-Speed Photography Technology

2.1 Introduction

In this chapter, the development history of the high-speed photography technology is introduced, as well as its application scenarios in experimental fluid mechanics. Then, some typical high-speed photography experimental systems are introduced, including platforms achieved by combining high-speed photography with schlieren method and other technologies. Finally, the future prospects of the high-speed photography technology are discussed.

2.2 High-Speed Imaging

High-speed photography is a specialized photographic technique that captures the instantaneous image of an object. Frame rates of ordinary cameras are typically 30 or 60 frames per second (fps). In this book, photography with a frame rate exceeding 100 fps is considered as the high-speed photography. High-speed imaging technology has always been highly praised because it has the capability to capture phenomena that are difficult to observe with the eyes. Recently, high-speed technology has been widely applied in the fluid mechanics, playing a crucial role in exploring the physical mechanisms in fluid. Subsequently, the history, camera types, and applications of the high-speed photography are introduced.

2.2.1 Development History

In 1868, Marey [1] employed a photographic rifle to record the flight of the birds and the movement of the insect wings. In 1878, Muybridge [2] constructed a system comprising 12 cameras in order to investigate whether a horse lifts all hooves off the ground while galloping, achieving the purpose of capturing high-speed motion by triggering them consecutively within a short period. Based on this method, Brixner [3] utilized 37 ordinary cameras, spaced 1 ms apart, to capture a nuclear explosion.

In 1936, Edgerton [4] invented the stroboscopic flash, which could be applied for the high-speed still photography. This technique utilizes microsecond flashes to freeze the motion of the objects, enabling him to observe many common motions and fluid phenomena [5, 6]. Specifically, this technique involves capturing only one image per repeatable event over multiple events, varying the timing of each image to achieve the high-speed photography. Consequently, repeatable events such as the inkjet drop ejection and the water drop impact can be observed and investigated effectively. Worthington [7] extensively explored the splash motion on solids and fluids due to impact through stroboscopic imaging techniques. When a sphere or drop falls into the water, the cavity that collapses under the water pressure is named the Worthington jet. Until now, stroboscopic imaging technology is still applied in the visualization of the high-speed flows due to its advantages of the low cost, high resolution, and superior dynamic range.

In 1949, Miller [8] introduced the rotating mirror technique, significantly enhancing the speed of the high-speed imaging. This technique, based on the schlieren method, records continuous images onto a circular recording surface through a rotating relay lens. In the 1950s, this technology was applied to the research on the thermonuclear weapons, achieving 1 million frames per second (Mfps). Subsequently, Chin et al. [9] developed it to a shooting speed of 25 Mfps. Subsequent advancements led to the development of the rotating prisms, rotating mirrors, and rotating drum cameras.

Later, in pursuit of faster shooting speeds, the photoelectric camera system achieved significant advancements. Specifically, with the advancement of the optoelectronic technology, image sensors such as CCD (Charge-Coupled Device) and CMOS (Complementary Metal–Oxide–Semiconductor) have gradually replaced the photosensitive film [10, 11], leading to the development of the digital imaging technology. For instance, image converter cameras convert incident photons into electron beams before recording the image, which are then accelerated and focused onto a phosphor screen [12]. Currently, the speed of the digital high-speed imaging technology ranges from 200 fps to 200 Mfps.

2.2.2 Application Scenarios

The application scenarios of the high-speed photography in scientific research includes extensive fields such as physics, chemistry, biology, and materials science.

In physics research, high-speed photography is often applied to capture particle motions such as electrons and photons within extremely short periods of time, exploring the mechanisms of their interactions [13, 14]. Furthermore, high-speed photography is also applied to investigate the flow characteristics of the fluids, including dynamic characteristics of the cavitation, drop, and wake [15].

In chemistry research, high-speed photography is utilized to explore the mechanisms of rapid reactions such as explosions and combustion [16, 17]. Combined with the spectral analysis, high-speed photography can also be employed to investigate the molecular structural characteristics [18].

In biology research, high-speed photography is employed to observe animals' movements such as flight and jumping, allowing for the exploration of their physiological characteristics [19]. At the microscopic level, processes such as cell division often occur within extremely short periods of time, which can be observed through the microscopes and the high-speed photography [20, 21].

In materials science research, the deformation and fracture of the materials are also high-speed processes, and the high-speed photography can capture these instantaneous phenomena, providing insights into the material properties and stress conditions [22].

2.3 Typical High-Speed Photography Experimental System

In this section, the configuration of typical high-speed photography experimental platform in fluid mechanics is presented. Its combined applications with methods, such as schlieren method and particle image velocimetry (PIV), are also exhibited.

2.3.1 Visualization Platform for Bubble Oscillation

Visualization platform for the bubble oscillation is a typical high-speed photography experimental system, characterized by its simple structure and ease of operation. Notably, its modular design allows for the convenient substitution of the research object with other objects, such as the drops and particles.

The bubble generation method can be categorized into two types: laser-induced and electrical spark-induced. Here, the laser-induced cavitation bubble experimental setup is presented. Figure 2.1 presents the visualization platform for the bubble oscillation. The experimental platform consists of a laser generation module (laser generator), a high-speed photography module (high-speed camera), a control module (delay generator), an optical component module, an illumination module, a motion control module (three-dimensional translation stage and experimental objects), a pressure monitoring module (hydrophone and oscilloscope), etc., which are all mounted on a vibration isolation platform [23]. Among them, the motion control module and laser generation module are unique to the cavitation bubble research. For research on drop and particle motion, these modules can generally be replaced directly. Furthermore, based on this experimental platform, it is convenient to incorporate technologies such as Particle Image Velocimetry (PIV) for better observation

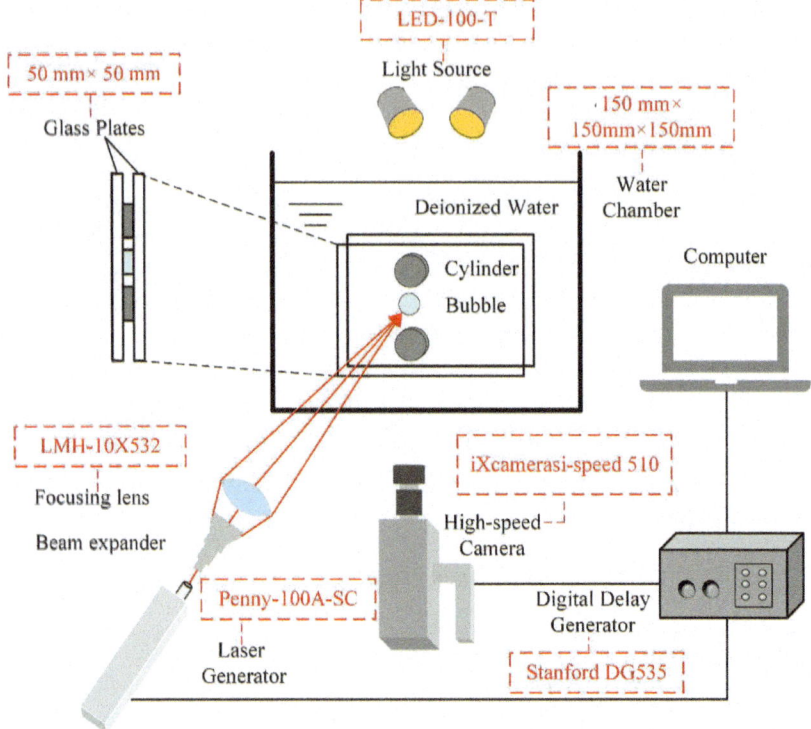

Fig. 2.1 The configuration of the visualization platform for the bubble oscillation. Reprinted with the permission from Ref. [23] Copyright (2024) (AIP Publishing)

of the fluid or cavitation motion characteristics. Specifically, PIV is a technique commonly applied to measure the flow field characteristics. It involves adding tracer particles to the fluid and recording their motion trajectories to obtain the velocity distribution [24].

2.3.2 Schlieren Method Imaging Platform

When light passes through the measured flow field, the light refractive index varies accordingly due to variations in air density within the flow field, causing the light to deflect. Based on this, the schlieren method is developed and widely applied in observing phenomena such as shock waves and airflow boundary layers [25].

Figure 2.2 shows the configuration of the schlieren method imaging platform. In the figure, the schlieren setup images the scene onto a slit plane through a flash light. Subsequently, a series of lenses are arranged to create a collimated light path. The parallel beam of light passes through the experimental object, and is then blocked

2.3 Typical High-Speed Photography Experimental System

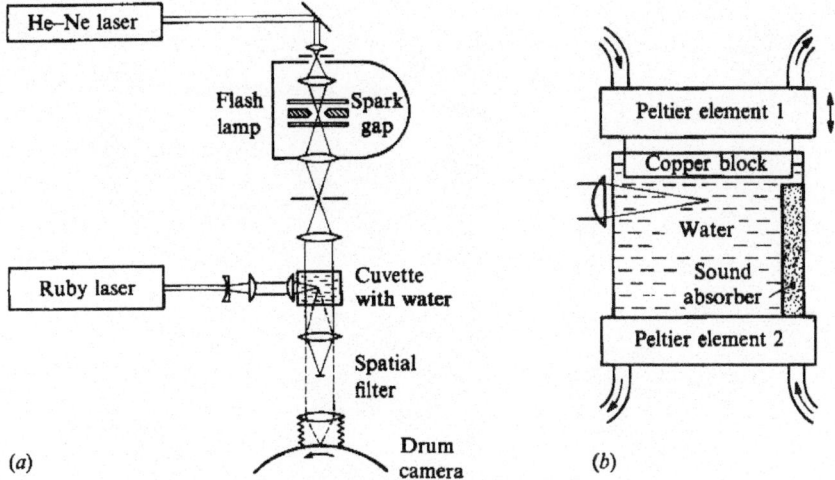

Fig. 2.2 The configuration of the schlieren method imaging platform. Reprinted with the permission from Ref. [27] Copyright (1989) (Cambridge University Press)

by a vertical wire. At this point, half of the light is cut off, creating a schlieren image with alternating light and dark areas. The schlieren method can be also implemented through the knife-edge set in the vertical position [26].

2.3.3 Water/Wind Tunnel Platform

In the research on the cavitation flow, the water tunnel platform generally consists of a drive module, a control module, a measurement module, and a research module [28]. The platform provides water circulation through a water pump, and regulates experimental conditions, including the flow velocity and pressure. The research module includes an observation window and the high-speed photography section, which are employed to observe the cavitation phenomena and flow states in the experiments. Figure 2.3 shows the configuration of the water tunnel platform. Objects 1 to 9 are water tank, shrink section, test section, diffusion section, electromagnetic flowmeter, pump, pressure sensor, high-speed camera, and S-shaped hydrofoil, respectively.

For the wind tunnel platform, it achieves the stable airflow through the air compressor (or fan), with the remaining components being similar to those of a water tunnel platform [29]. Additionally, in order to simulate complex airflow conditions, the wind tunnel requires a control system that can adjusts wind speed and direction.

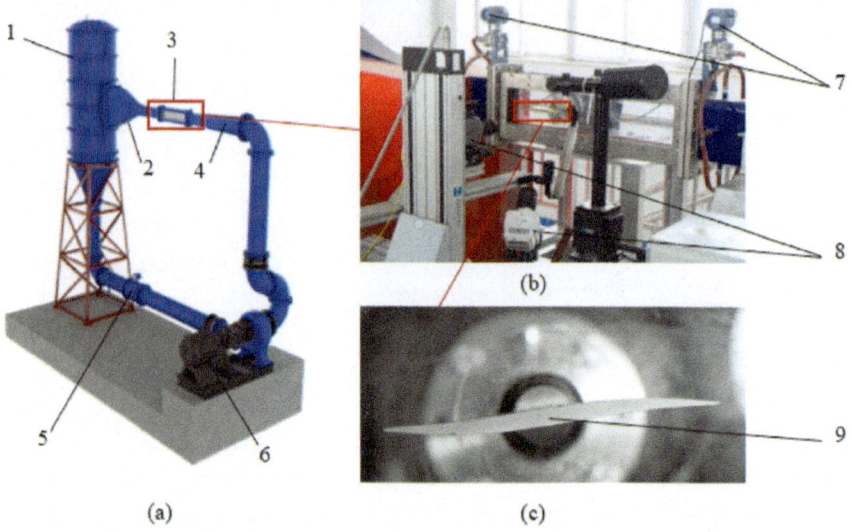

Fig. 2.3 The configuration of the water tunnel platform. Reprinted with the permission from Ref. [28] Copyright (2023) (ELSEVIER)

2.4 Future Directions

The future development of the high-speed photography technology lies in achieving faster speeds, clearer images, and the ability to capture microscopic objects. High-speed digital holography enables the digital reconstruction of the holograms by simulating the optical diffraction process through computers [30]. This method can record and recreate high-resolution images of the objects, making the details of the images even clearer. It also preserves the full amplitude and phase information of the light waves, allowing for the recreation of the three-dimensional characteristics. Furthermore, for the high-speed observation techniques at the nanoscale, the scanning probe microscopy can be applied [31]. It utilizes a nanometer-scale three-dimensional displacement positioning control system and an extremely sharp probe to observe the surface characteristics of nanoscale objects.

References

1. Marey EJ (1868) Determination experimentale du mouvement des ailes des insectes pendant le vol. CR Acad Sci Paris 67:1341–1345
2. Muybridge E, Mozley AV (1887) Human and animal locomotion. Dover, New York
3. Brixner B (1992) In: Dewey JM, Racca RG (eds) Proceedings of 20th international congress on high speed photography and photonics, vol 1801. SPIE, Bellingham, WA, pp 52–60
4. Versluis M (2013) High-speed imaging in fluids. Exp Fluids 54:1–35
5. Edgerton HE, Killian JR (1954) Flash!: Seeing the unseen by ultra high-speed photography

6. Edgerton HE, Jussim E, Kayafas G (1987) Stopping time: the photographs of Harold Edgerton
7. Worthington AM (1908) A study of splashes. Longmans, Green, and Company
8. Miller CD (1949) Half-million stationary images per second with refocused revolving beams. J Soc Motion Picture Eng 53(5):479–488
9. Chin CT, Lancée C, Borsboom J et al (2003) Brandaris 128: a digital 25 million frames per second camera with 128 highly sensitive frames. Rev Sci Instrum 74(12):5026–5034
10. Boyle WS, Smith GE (1970) Charge coupled semiconductor devices. Bell Syst Tech J 49(4):587–593
11. Sangster FLJ, Teer K (1969) Bucket-brigade electronics: new possibilities for delay, time-axis conversion, and scanning. IEEE J Solid-State Circuits 4(3):131–136
12. Garfield BRC (1903) Developments in image converter streak/framing camera systems. In: 20th international congress on high speed photography and photonics. SPIE 1801:192–203
13. Huston AE (1978) High-speed photography and photonic recording. J Phys E: Sci Instrum 11(7):601
14. Huston AE, Walters F (1962) Electron tubes for high-speed photography. Adv Electron Electron Phys. Academic Press 16:249–263
15. Thoroddsen ST, Etoh TG, Takehara K (2008) High-speed imaging of drops and bubbles. Annu Rev Fluid Mech 40(1):257–285
16. Cairns RW (1944) Study of high explosives by high-speed photography. Ind Eng Chem 36(1):79–85
17. Mao G, Shi K, Zhang C et al (2020) Experimental research on effects of biodiesel fuel combustion flame temperature on NOX formation based on endoscope high-speed photography. J Energy Inst 93(4):1399–1410
18. Ando T, Uchihashi T, Scheuring S (2014) Filming biomolecular processes by high-speed atomic force microscopy. Chem Rev 114(6):3120–3188
19. Straw AD (2021) Review of methods for animal videography using camera systems that automatically move to follow the animal. Integr Comp Biol 61(3):917–925
20. Mikami H, Lei C, Nitta N et al (2018) High-speed imaging meets single-cell analysis. Chem 4(10):2278–2300
21. Lau AKS, Tang AHL, Xu J et al (2015) Optical time stretch for high-speed and high-throughput imaging—from single-cell to tissue-wide scales. IEEE J Sel Top Quantum Electron 22(4):89–103
22. Kirugulige MS, Tippur HV, Denney TS (2007) Measurement of transient deformations using digital image correlation method and high-speed photography: application to dynamic fracture. Appl Opt 46(22):5083–5096
23. Shen J, Ying J, Liu W et al (2024) The evolution of the bubble collapse morphology between two cylinders within a confined space. Phys Fluids 36(10)
24. Verhaagen B, Boutsioukis C, Heijnen GL et al (2012) Role of the confinement of a root canal on jet impingement during endodontic irrigation. Exp Fluids 53:1841–1853
25. Hargather MJ, Settles GS (2012) A comparison of three quantitative schlieren techniques. Opt Lasers Eng 50(1):8–17
26. Požar T, Agrež V (2021) Laser-induced cavitation bubbles and shock waves in water near a concave surface. Ultrason Sonochem 73:105456
27. Vogel A, Lauterborn W, Timm R (1989) Optical and acoustic investigations of the dynamics of laser-produced cavitation bubbles near a solid boundary. J Fluid Mech 206:299–338
28. Liu H, Guo Q, Shi L et al (2023) Lift-drag characteristics of S-shaped hydrofoil under different cloud cavitation conditions. Ocean Eng 278:114374
29. Shen P, Lin L, Wei Y et al (2019) Vortex shedding characteristics around a circular cylinder with flexible film. Eur J Mech-B/Fluids 77:201–210
30. Kakue T, Endo Y, Nishitsuji T et al (2017) Digital holographic high-speed 3D imaging for the vibrometry of fast-occurring phenomena. Sci Rep 7(1):10413
31. Tuma T, Lygeros J, Kartik V et al (2012) High-speed multiresolution scanning probe microscopy based on Lissajous scan trajectories. Nanotechnology 23(18):185501

Chapter 3
Visualization Research on Bubble Dynamics

3.1 Introduction

In this chapter, the bubble behaviors in various scenarios are introduced. Firstly, the bubble morphological evolution, collapse jet, shock wave and other characteristics near various boundaries are analyzed. Secondly, the impact mechanisms of the particle motion are explored. Finally, the flow characteristics and structures under various cavitating flow patterns are introduced and demonstrated in detail.

In the visualization research on the bubble dynamics, the speeds of the high-speed photography are mostly below 300,000 fps. And, multi perspective shooting is applied, such as observing the evolution of the collapse jet from multiple perspectives. When it comes to the phenomena that are invisible by normal photography techniques (e.g., shock waves), the schlieren technique is applied in the experiments. At this point, high-speed photography techniques at 100 million fps are employed to observe the shock waves. In addition, instability waves on the bubble interface can also be observed through schlieren technique, which is brighter due to its high scattering efficiency.

3.2 Spherical Bubble Near Various Boundaries

This section introduces the typical collapse phenomena of the bubbles near different rigid walls, including bubble deformation, collapse jets, and shock waves. In order to demonstrate the evolutionary characteristics of these phenomena, high-speed photography experimental techniques with different speeds are employed, with a maximum rate of up to 100,000,000 fps.

3.2.1 Bubble Deformation

When the bubble oscillates freely, it only manifests as the bubble interface movement in the radial direction. Rigid walls always exert attractive influence on the bubble, manifesting as anisotropic bubble motion towards the wall. The maximum bubble interface velocity is generally on the order of 10 m/s, and a speed of less than 300,000 fps for observing the bubble deformation is sufficient.

Figure 3.1 shows the photographs of a bubble oscillating near a flat wall at 300,000 fps. In the figure, the bubble undergoes significant deformations. A depression is observed at the interface far from the wall, and the bubble moves towards the wall, which demonstrate the attractive effect of the wall.

As the boundary becomes more complex, the bubble exhibits more complex deformation characteristics. Figure 3.2 shows the photographs of a bubble collapsing in a rectangular channel at 100,000 fps. Subfigures (a)–(c) correspond to large, medium, and small distances between the bubble and the sidewall, respectively. In subfigure (a), as the bubble reaches maximum volume, its upper and lower interfaces are restricted by the channel and remain in contact during the subsequent oscillation process. In the initial collapse stage, a symmetric annular depression is generated in the middle. And, the bubble collapses in a typical hourglass shape. Subsequently, the bubble splits into two parts, upper and lower, and attaches to the sidewalls of the channel, connected by a transparent tubular structure in the middle. During the rebound process, the two parts of the bubble and the structure in the middle continue to grow.

As shown in subfigure (b), the bubble-sidewall distance is medium. During the collapse process, the annular depression is still observed, but it is not symmetric. The bubble depression closer to the sidewall becomes smaller, and the depression far from the sidewall becomes larger. Subsequently, the transparent tubular structure is closer to the sidewall. In subfigure (c), the bubble-sidewall distance is the smallest. The depression is only observed on the side far from the sidewall. At this point, the bubble collapses in a crescent shape and does not split during the collapse process.

Figure 3.3 illustrates the photographs of a bubble oscillating near a convex wall at 30,000 fps. Subfigures (a)–(c) correspond to small, medium, and large distances between the bubble and the convex wall, respectively. In subfigure (a), during the growth process, the bubble experiences significant restriction from the wall. During

Fig. 3.1 High-speed photography of a bubble oscillating near a flat wall at 300,000 fps. Reprinted with the permission from Ref. [1] Copyright (1975) (Cambridge University Press)

3.2 Spherical Bubble Near Various Boundaries

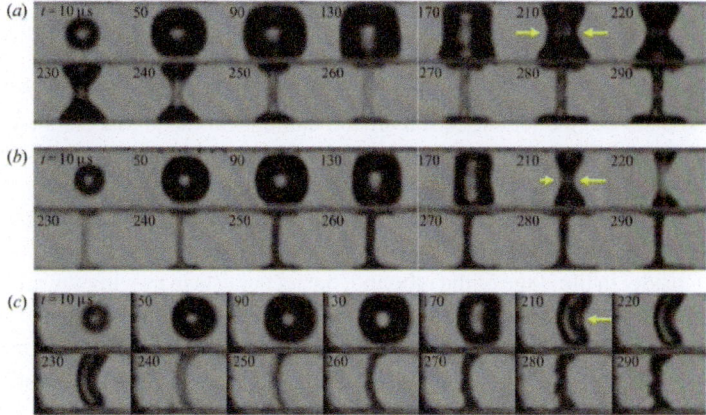

Fig. 3.2 High-speed photography of a bubble collapsing in a rectangular channel at 100,000 fps. Reprinted with the permission from Ref. [2] Copyright (2022) (Cambridge University Press)

the collapse process, a neck forms at the point where the bubble is in contact with the wall. Subsequently, the bubble collapses in a typical mushroom shape. As the bubble-wall distance increases (i.e., subfigure (b)), the bubble-wall contact area is relatively small during the growth process. The neck is not observed, and the bubble centroid moves towards the wall. In the rebound stage, most of the microbubble clouds adhere to the wall. In subfigure (c), in the growth stage, the bubble is not in contact with the wall, which then presents an elliptical shape.

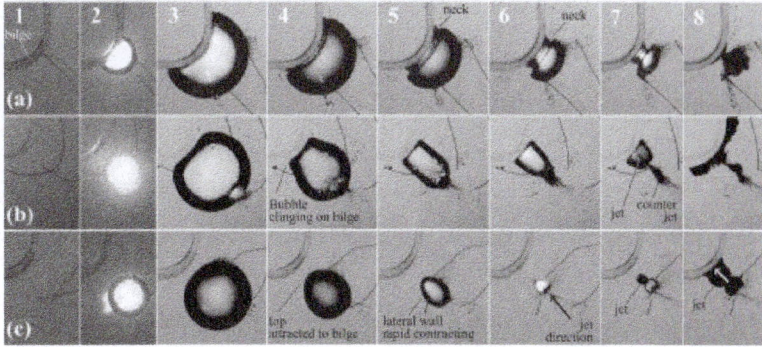

Fig. 3.3 High-speed photography of a bubble collapsing near a convex wall at 30,000 fps. Reprinted with the permission from Ref. [3] Copyright (2013) (ELSEVIER)

3.2.2 Collapse Jet

At an appropriate bubble-wall distance, the collapse jet can be observed. The different types of the jets and the jet deflection are the main phenomena that can be observed near rigid boundaries.

Figure 3.4 illustrates the photographs of the jet and counter-jet near a flat wall at 20,000 fps. Subfigures (a) and (b) illustrate the cases without and with counter-jet, respectively. As shown in subfigure (a), a jet that penetrates the bubble can be observed inside. In the stage of the rebound, the jet persists in moving towards the flat wall, shaping a prominent pointed structure at the bottom. This collapse jet drives the bubble to move towards the wall. At a small bubble-wall distance (i.e., subfigure (b)), a counter-jet is observed and drives the bubble away from the wall in the subsequent process [4].

In addition, through multi perspective shooting techniques, the formation and evolution process of the collapse jet can be further explored. Figure 3.5 illustrates the high-speed photographs of the side and top views of the collapse jet near a flat wall at 20,000 fps. Subfigures (a) and (b) illustrate the side view and top view, respectively. As shown in subfigure (a), during the third to fifth bubble oscillation periods, the bubble volume becomes very small, and the collapse phenomena cannot be well observed in a side view. In subfigure (b), one can observe interesting phenomena from a top view. During the second period, the jet is formed at top center of the bubble and maintains until the end. During the third to fifth bubble oscillation periods, the bubble oscillates in an annular shape, and its radius continuously increases.

When a bubble oscillates, instability waves are generated on its interface, which can be observed through the schlieren technique. Figure 3.6 shows the high-speed photography taken with schlieren technique at 20,000 fps. In the figure, some areas

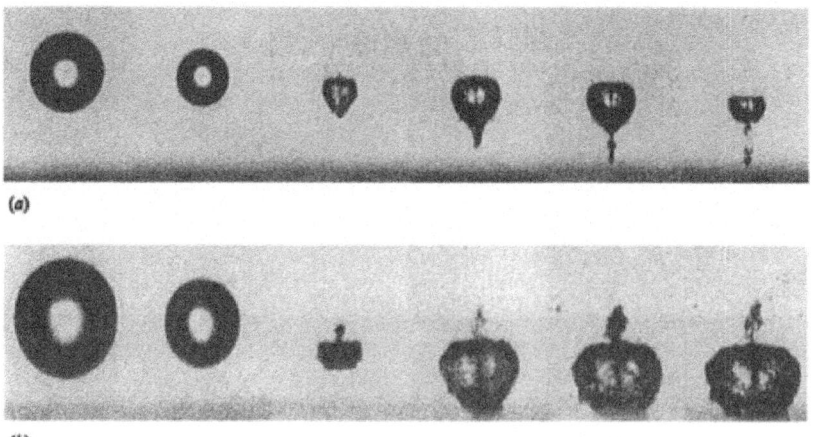

Fig. 3.4 High-speed photography of the jet and counterjet near a flat wall at 20,000 fps. Reprinted with the permission from Ref. [5] Copyright (1989) (Cambridge University Press)

3.2 Spherical Bubble Near Various Boundaries

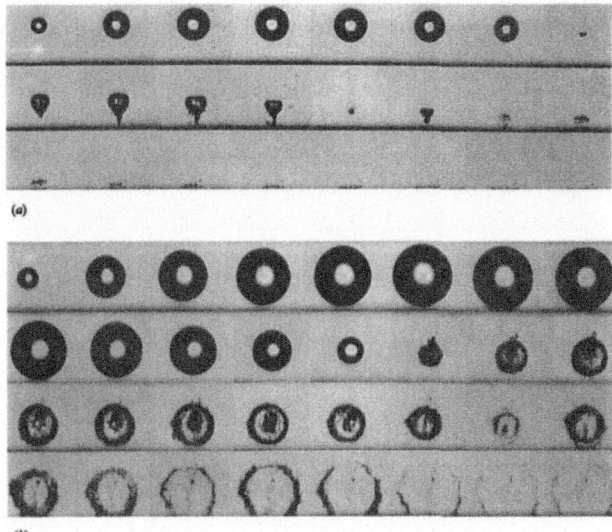

Fig. 3.5 High-speed photography of the side and top views of the collapse jet near a flat wall at 20,000 fps. Reprinted with the permission from Ref. [5] Copyright (1989) (Cambridge University Press)

are white due to their high scattering efficiency, namely the cloud of microbubbles on the bubble interface. The bubble is only brighter on the interface during the first oscillation period. Especially the depression part on the upper bubble interface. During the second bubble oscillation period, the counter-jet is the brightest, which is actually composed of the cloud of microbubbles.

Usually, the collapse jet moves towards the rigid wall due to its attractive effect. When there are multiple walls or a complex wall near the bubble, the direction of the

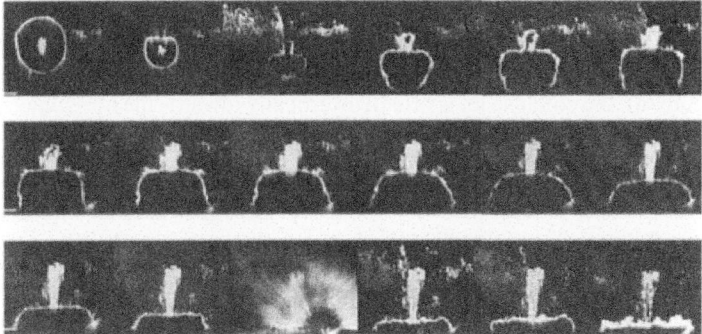

Fig. 3.6 High-speed photography of a bubble oscillating near a flat wall taken with schlieren technique at 20,000 fps. Reprinted with the permission from Ref. [5] Copyright (1989) (Cambridge University Press)

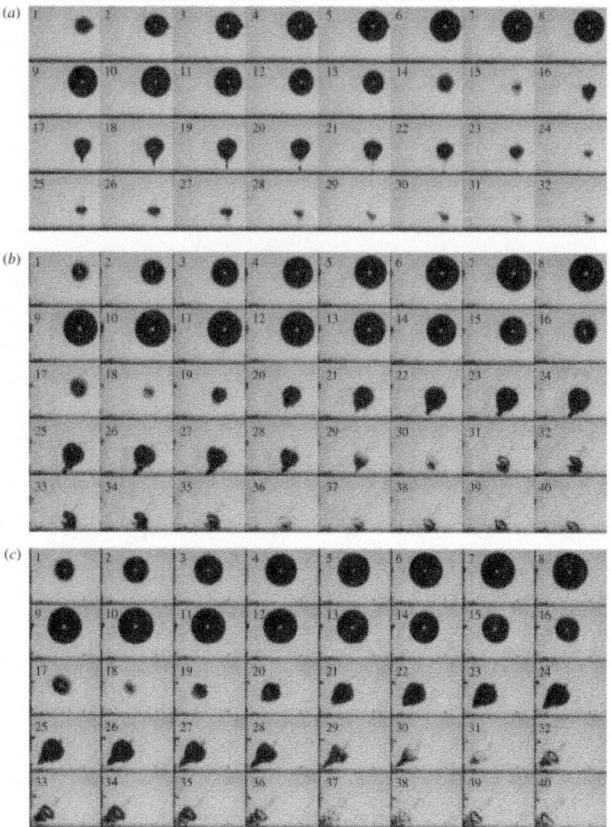

Fig. 3.7 High-speed photography of a bubble collapsing near a right-angled wall at 100,000 fps. Reprinted with the permission from Ref. [6] Copyright (2018) (Cambridge University Press)

jet undergoes significant variations at different positions. Figure 3.7 illustrates the photographs of a bubble oscillating near a right-angled wall at 100,000 fps. Subfigures (a)–(c) correspond to large, medium, and small bubble-wall distances, respectively. In subfigure (a), the collapse jet is oriented vertically towards the horizontal wall. As the distance decreases, the jet direction is increasingly inclined towards the vertical wall. Thus, the distance significantly affects the jet direction.

3.2.3 Shock Wave

Shock waves are generally difficult to observe through normal photography techniques, which require the schlieren technique. Due to the high velocity of the shock waves, extremely high speeds of the camera are required to capture the image.

3.2 Spherical Bubble Near Various Boundaries

Figure 3.8 illustrates the photographs of the shock waves near a flat wall at 100,000,000 fps. In the figure, during the bubble collapse process, multiple shock waves are generated sequentially. Firstly, two weak shock waves are observed at two ends of the bubble, due to the contact between the jet and the lower bubble interface [7]. Subsequently, the jet forms a tip at its lower part, where a strong shock wave is generated.

Based on the schlieren technique, Požar et al. [8] captured clearer images of the shock wave evolution. Figure 3.9 illustrates the photographs of the shock waves near a concave wall at 210,000 fps. At the initial moments of three oscillation periods, strong shock waves are generated. During the evolution of these three shock waves, they interact with each other and form multiple refocusing points (i.e., B4, D3, and D4 in Fig. 3.9).

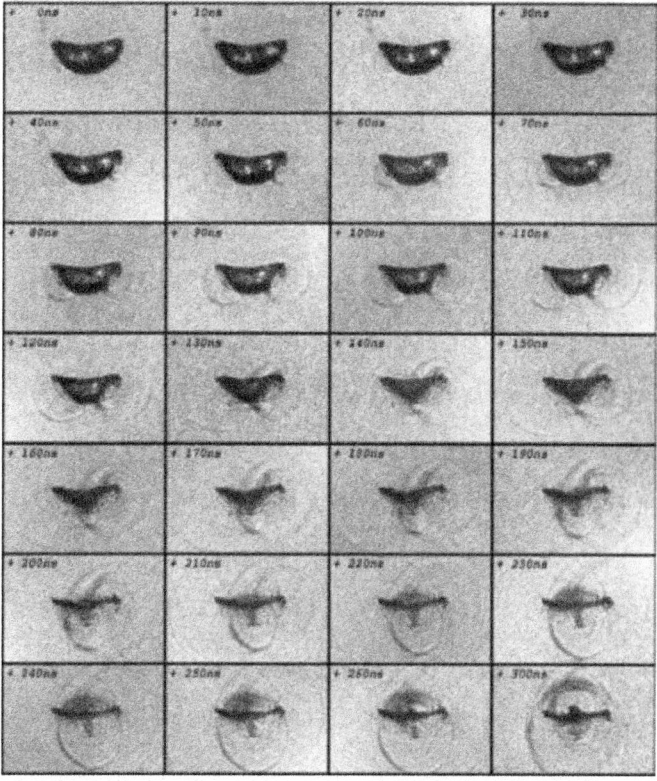

Fig. 3.8 High-speed photography of the shock waves near a flat wall at 100,000,000 fps. Reprinted with the permission from Ref. [7] Copyright (2003) (Cambridge University Press)

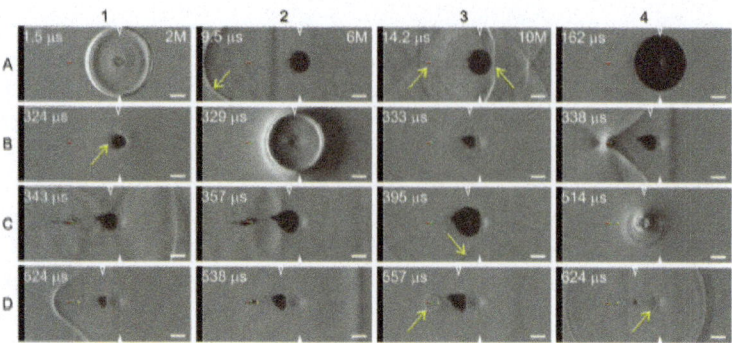

Fig. 3.9 High-speed photography of the shock waves near a concave wall at 210,000 fps. Reprinted with the permission from Ref. [8] Copyright (2021) (ELSEVIER)

3.3 Bubble Within Narrow Gaps

This section introduces the typical collapse phenomena of the bubbles within narrow gaps. In these scenarios, the bubbles are restricted by the narrow gap and only exhibit radial motion, with negligible axial motion. At this point, the bubble takes on a cylindrical shape, and its characteristics of the morphological evolution and collapse jet are significantly different from those of a spherical bubble. The time scale of a bubble collapsing within a narrow gap is similar to that of a spherical bubble, so the camera speed is usually within the same range. In addition, since the bubbles within the narrow gap only exhibit radial motion, adding micro particles to the fluid can demonstrate the velocity distribution through the particle image velocimetry (PIV) algorithm.

3.3.1 Flat Boundary

The jet of a spherical bubble is generally composed of the cloud of microbubbles, which penetrates the bubble from the inside and cause significant deformation on its surface. When it comes to the jet of a bubble within the narrow gap, due to the limitation of the axial space, it is typically a pure liquid jet that divides the bubble into multiple segments.

Figure 3.10 illustrates the photographs of the bubble oscillating near a rigid wall within a narrow gap at 1,000,000 fps. In the figure, a large number of micro particles are placed in the liquid. Subfigure (a) shows the first oscillation period. Subfigure (b) and (c) illustrate the liquid velocity distributions. In subfigure (a), the bubble is transparent and the surrounding interface is darker, forming a typical cylindrical shape. In the growth stage, the bubble maintains a good circular shape. In the initial collapse stage, the bubble interface on the left and right sides away from the rigid wall first contract. Meanwhile, the lower bubble interface adheres to the wall. At

3.3 Bubble Within Narrow Gaps

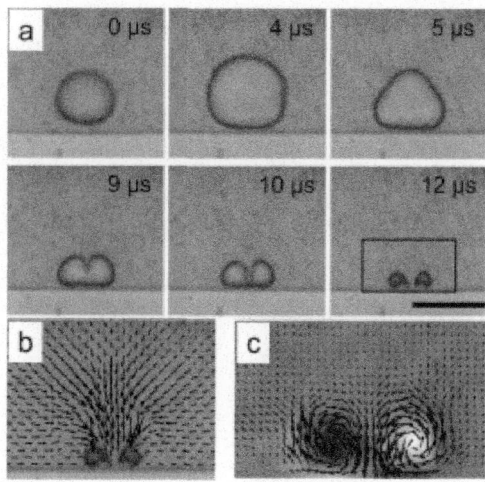

Fig. 3.10 High-speed photography of a bubble collapsing near a rigid wall within a narrow gap at 1,000,000 fps. Reprinted with the permission from Ref. [9] Copyright (2007) (American Physical Society)

this point, the bubble obtains a triangular shape. Subsequently, the upper interface rapidly contracts downward and develops into a liquid jet, pointing vertically towards the wall. Then, a liquid jet pierces through the bubble, causing it to divide into two distinct sections, continuing to move along both sides of the wall.

Based on the PIV algorithm, the velocity distribution at the end of the collapse is exhibited (i.e., subfigure (b)). It can be observed that the liquid velocity is highest in the areas where the interface depression occurs and the liquid jet flows through. In subfigure (c), two parts of the bubble vanish and two vortex rings are generated, with high liquid velocities.

The free surface can also be considered as a type of the flat boundary. An oscillating bubble has a significant impact on the free surface, and its collapse behavior differs from that of the rigid boundary. Figure 3.11 illustrates the photographs of a bubble oscillating near a free surface within a narrow gap at 30,000 fps. In the figure, as the bubble reaches maximum radius (i.e., 0.82 ms), the free surface is disturbed and protrudes upward. During the collapse process, a section of free surface is drawn towards the bubble and moves downward, and another portion continues to move upward. Subsequently, the bubble moves downwards and forms a depression at the top. As the collapse progresses, the depression develops into a downward jet, which then splits the bubble into two parts. These two parts of the bubble rotate in opposite directions. Meanwhile, the concave portion of the free surface also forms an upward jet. Throughout this period, the spray continues to elongate upward.

3.3.2 Curved Boundary

When it comes to the curved boundaries, the bubble exhibits more complex collapse phenomena. Figure 3.12 illustrates the photographs of the bubble oscillating near

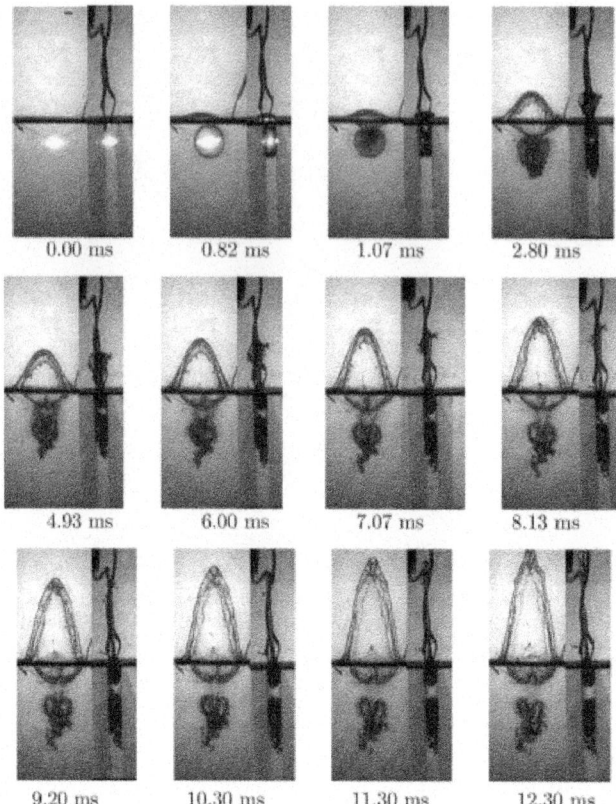

Fig. 3.11 High-speed photography of a bubble oscillating near a free surface within a narrow gap at 30,000 fps. Reprinted with the permission from Ref. [10] Copyright (2018) (ELSEVIER)

double cylinders within a narrow gap at 100,000 fps. In the figure, two liquid films are observed on the surface. It is due to the high growth velocity of the bubble, and the liquid between the bubble and the cylinders is compressed, resulting in the liquid films on the axial bubble surface. The liquid films move inward and form protrusions on the left and right interfaces. The bubble centroid of the subsequent collapse moves horizontally to the left.

Cylindrical drops can also be regarded as a type of cylindrical boundary. Figure 3.13 shows the photographs of a bubble oscillating inside a cylindrical drop within a narrow gap at 100,000 fps. Subfigures (a), (c), and (e) correspond to the bubble oscillation process during the first period. Subfigures (b), (d), and (f) are the partial enlarged images. In the figure, the large and small circles represent the cylindrical drop and the bubble, respectively. During the growth process, the radius of the drop grows and a ring-shaped jet is observed (i.e., 400 μs). As the bubble collapses, significant disturbances and jets are formed on the drop surface. In the initial stage of

3.4 Particle-Bubble Interaction

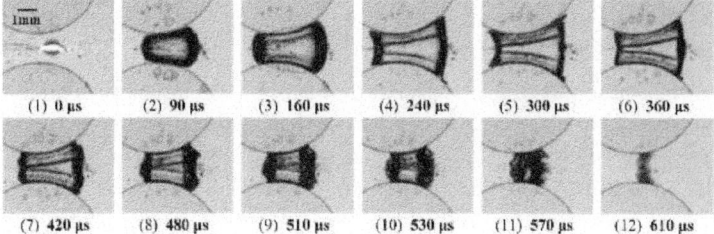

Fig. 3.12 High-speed photography of a bubble oscillating near double cylinders within a narrow gap at 100,000 fps. Reprinted with the permission from Ref. [11] Open access under a CC BY 4.0 license, https://creativecommons.org/licenses/by/4.0/

Fig. 3.13 High-speed photography of a bubble collapsing inside a cylindrical drop within a narrow gap at 100,000 fps. Reprinted with the permission from Ref. [12] Open access under a CC BY 4.0 license, https://creativecommons.org/licenses/by/4.0/

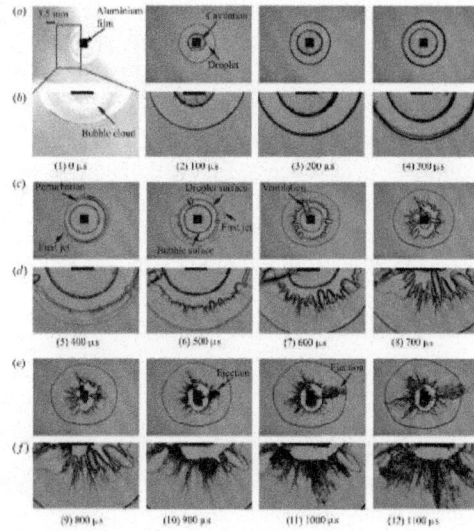

the rebound (i.e., 800 μs), a pressure impulse is generated, causing the liquid surface to move in the opposite direction.

3.4 Particle-Bubble Interaction

When the bubble oscillates near the particle, it can cause complex dynamic effects on the particle. At a small particle-bubble distance, the rapid expansion of the bubble and the collapse jet both accelerate the particle. When the distance is large, the bubble has an attractive effect on the particle. The shear force generated during the bubble growth process can also cause the rotation of the particle.

3.4.1 Single Particle

Usually, when it comes to the single particle motion, its radius is relatively small. Therefore, as the bubble oscillates, the particle has sufficient momentum and moves at a high velocity, with the speed of the camera generally reaching 1–10 million fps.

Figure 3.14 shows the photographs of a particle accelerated by a bubble at 10,000,000 fps. Subfigures (a) and (b) are two typical cases of the explosively expanding cavitation bubbles. In the figure, the particle settles freely from the top with a radius of approximately 30 μm. As shown in subfigure (a), three sets of the particle-bubble interactions are demonstrated. It can be observed that the particles move away from the bubble. In subfigure (b), the process of a particle detaching from the bubble is demonstrated. The bubble radius is 150 μm at 8 μs. At this moment, the particle is about to detach from the bubble, and it is connected to the bubble by a necklike structure. Subsequently, the particle detaches and the bubble continues to grow to its maximum radius (170 μm at 24.2 μs).

When a bubble collapses, it generally attracts the particles. Ren et al. [14] placed a particle (with a radius of 2.5 mm) on a rigid wall to restrict its movement during the bubble growth process. Figure 3.15 shows the photographs of a particle accelerated by a bubble at 1,000,000 fps. Subfigures (a)–(d) correspond to the bubbles located near the particle at different angles and distances. As shown in subfigure (a), at the moment of maximum bubble radius (i.e., 0.203 ms), the particle detaches from the wall. In the collapse stage, the particle does not show significant movement. After multiple rebounds of the bubble, the particle moves vertically upward, with a height

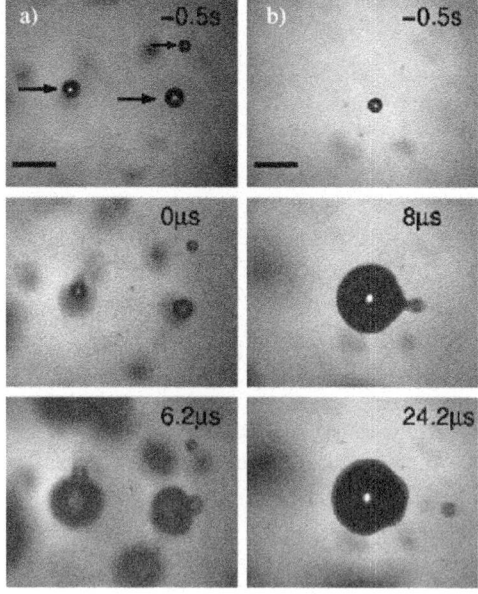

Fig. 3.14 High-speed photography of a particle accelerated by a bubble at 10,000,000 fps. Reprinted with the permission from Ref. [13] Copyright (2004) (American Physical Society).

3.4 Particle-Bubble Interaction

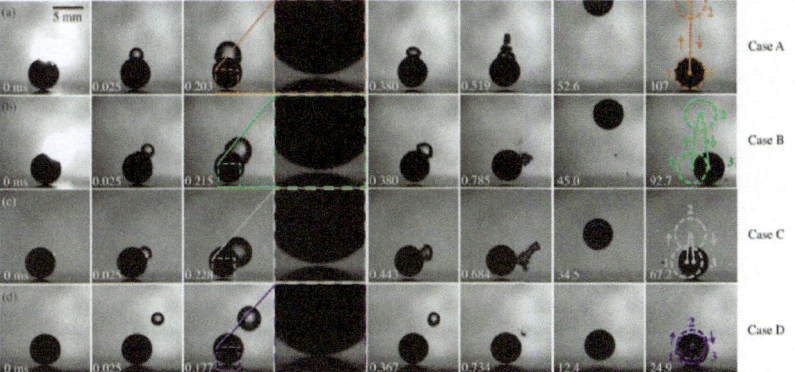

Fig. 3.15 High-speed photography of a particle accelerated by a bubble near a rigid wall at 1,000,000 fps. Reprinted with the permission from Ref. [14] Copyright (2022) (American Physical Society).

of 4.6 times the particle radius. Subsequently, the particle bounces back several times on the wall. Thus, the bubble collapses have a strong attraction effect on the particle.

When the position of the bubble deviates from the vertical direction (i.e., subfigure (b)), the particle still detaches from the wall during the bubble growth process. Subsequently, the particle moves in a parabolic motion away from the bubble. At this point, the particle motion can be decomposed into upward and leftward movements. The height of upward movement is smaller than that in subgraph (a). The leftward movement may be caused by the bubble collapse jet and the shock waves. When the deviation of the bubble position from the vertical direction is greater (subfigure (c)), the particle still detaches from the rigid wall during the bubble growth process, and its parabolic motion becomes weaker. As the particle-bubble distance increases, the particle detachment cannot be observed, and the parabolic motion of the particle becomes much weaker.

3.4.2 Multi-particle

Multiple particles can better demonstrate the repulsive and attractive effects of the bubble on them. Figure 3.16 shows the high-speed photography of the particles accelerated by a bubble at 50,000 fps. In the figure, the particle material is sand with the size of 75–106 μm, which settles freely from the top. The bubble is generated on a rigid wall. A particle is tracked and highlighted by the green boxes. In Fig. 3.16, the particles are significantly affected and move upward. As the bubble collapses, particles are attracted and move downward, at which point their speed reaches the maximum. At the end of the collapse, the particle-bubble distances are large enough, and the particles are less affected.

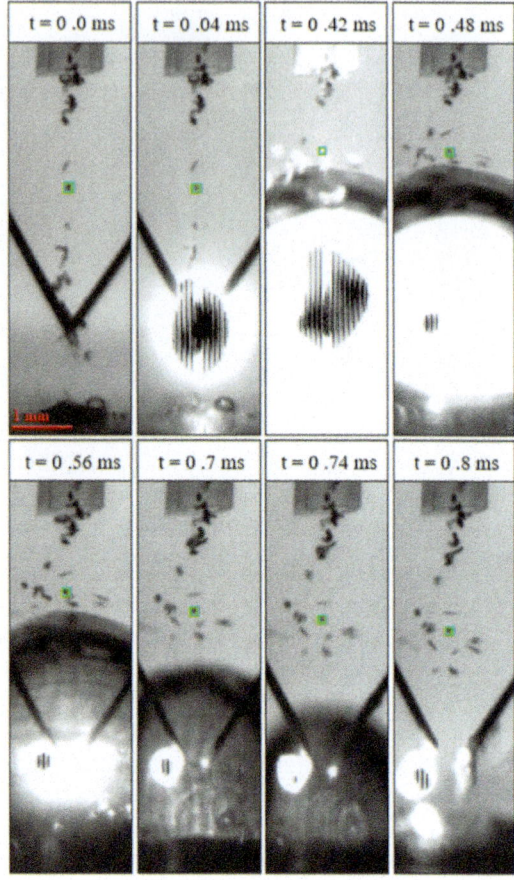

Fig. 3.16 High-speed photography of the particles accelerated by a bubble near a rigid wall at 50,000 fps. Reprinted with the permission from Ref. [15] Copyright (2018) (AIP Publishing).

When both particles and bubbles are located within the narrow gap, the bubbles only exhibit radial motion. At this point, the repulsive and attractive effects of the bubble on the particles can be observed more intuitively and clearly. Figure 3.17 illustrates the photographs of the particles accelerated by a bubble within a narrow gap at 36,000 fps. In the figure, the fluorescent particles (i.e., the bright objects) are placed in an annular shape on the substrate. At the moment of the bubble nucleation (0 μs), the particle distributions can be clearly observed due to electric spark discharge. During the bubble growth process (i.e., 56 μs), the particles move outward and forms a narrower ring. During the bubble collapse process (i.e., 195 μs), the particles move inward, exhibiting significant circumferential instability. Subsequently, the particles move outward again in the rebound stage. The reciprocating motion of the particles significantly deforms their annular shape.

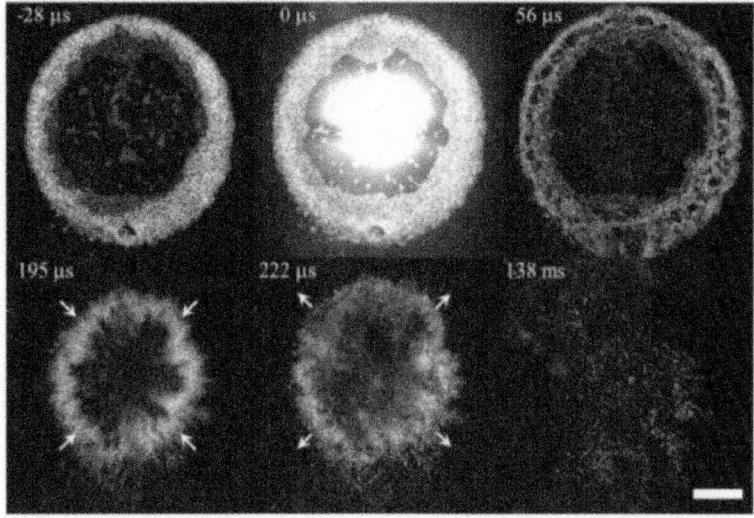

Fig. 3.17 High-speed photography of the particles accelerated by a bubble within a narrow gap at 36,000 fps. Reprinted with the permission from Ref. [16] Copyright (2020) (Cambridge University Press)

3.5 Cavitation Flow

This section discusses the cavitation flow in liquid machinery, namely attached cavitation. When the high-speed fluids pass over a solid surface, complex cavitation phenomena form on the solid surface. The cavitation flow characteristics are related to the surface shape, the flow angle, and the flow velocity. Once the surface shape and the flow angle are determined, the flow velocity becomes the decisive factor for the cavitation flow. Different velocities lead to different cavitation flows, and the higher the velocity, the more complex and unstable the cavitation area becomes. These cavitation flows move very quickly, and to discern the details, high-speed photography technology is employed to observe the phenomena.

3.5.1 Sheet Cavitation

Sheet cavitation typically occurs when a liquid experiences a pressure drop to the vapor pressure, leading to the formation of the cavitation along a solid surface. Sheet cavitation usually initiates at the leading edge of the propeller blades or hydrofoils, which is characterized by a thin vapor layer adhering to the surface. It is influenced by the local velocity and the pressure distribution around the object. Sheet cavitation is usually stable, but exhibits instability when encountering high-pressure areas at the trailing edge. This instability may lead to the shedding of the vapor structures

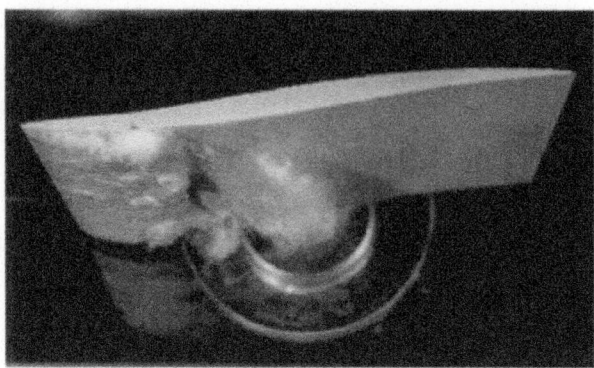

Fig. 3.18 High-speed photography of the sheet cavitation near the S-shaped hydrofoil at 7,000 fps. Reprinted with the permission from Ref. [17] Copyright (2023) (ELSEVIER)

and the formation of the cloud cavitation, which is a more turbulent and unstable cavitation flow.

Figure 3.18 illustrates the photographs of the sheet cavitation near the S-shaped hydrofoil. Cavitations are observed at the leading edge, indicating that the S-shaped hydrofoil is in the stage of the sheet cavitation. The fluid on the right side of the cavitation area becomes stable after a certain distance.

Figure 3.19 illustrates the photographs of the sheet cavitation near an axis-equipped hydrofoil and the cavitation within the gap. Subfigures A–D represent different gap spacings, respectively. As shown in subfigure A, the leading edge gap cavitation is almost entirely suppressed, and only an extremely small cavity is observed at the leading edge. In contrast, around the hydrofoil, a prominent sheet cavitation is visible near the leading edge. And, as it approaches the trailing edge, the cavitation becomes unstable, manifesting as smaller cloud cavitation. In subfigure B, with the increasing gap spacing, the cavity significantly expands, and the stability of the sheet cavitation increases, while cloud cavitation appears at the trailing edge. In subfigure C, as the gap spacing continues to increase, the leading edge sheet cavitation shifts towards the midsection of the hydrofoil, presenting a distinct sheet-like appearance. At the trailing edge, the cloud cavitation remains unchanged. As depicted in subfigure D, the gap spacing is the largest, the cavity at the leading edge expands, and the trailing edge gap exhibits pronounced cloud cavitation. The sheet cavitation in the midsection of the hydrofoil diminishes and transitions into more complex cloud cavitation.

3.5.2 Cloud Cavitation

As sheet cavitation transitions into a state of instability and detaches into the surrounding flow, it develops into the cloud cavitation. This progression results in

3.5 Cavitation Flow

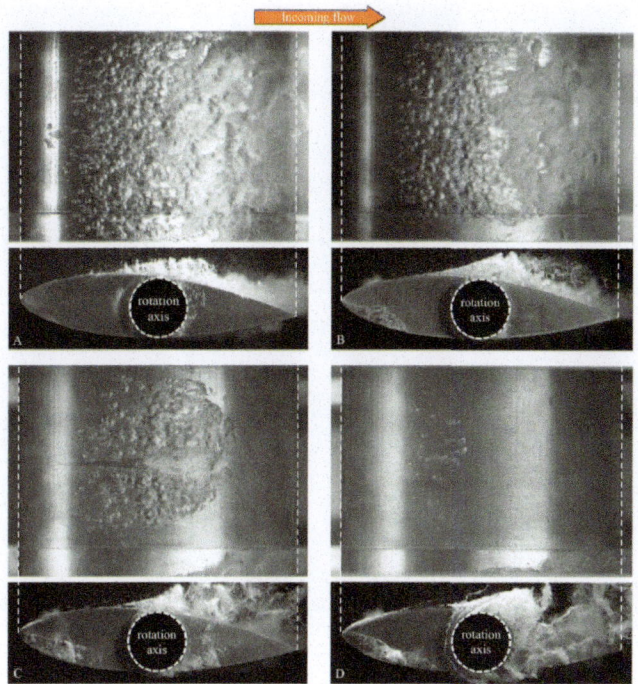

Fig. 3.19 High-speed photography of the sheet cavitation near an axis-equipped hydrofoil at 7,000 fps. Reprinted with the permission from Ref. [18] Copyright (2022) (ELSEVIER)

the vortex structure teeming with a multitude of the cavitation bubbles, creating a nebulous pattern. The cloud cavitation is characterized by the abundance of the micro-bubbles, appearing in regions of intense pressure fluctuations and swift flow dynamics, marking a phase of pronounced flow instability.

Figure 3.20 illustrates the photographs of the cloud cavitation near the mVG-2 hydrofoil at 10,000 fps. In subfigure (a), the "finger-like" vortex cavitation appears at the leading edge, with an initiation that is non-uniform in the spanwise direction. This "finger-like" vortex cavitation is a variant of the sheet cavitation. In subfigure (b), the "finger-like" cavitation evolves and extends downstream along with the primary flow. In subfigure (c), the "finger-like" cavitation begins to merge, forming smaller regions of the cloud cavitation, while the merged cavity surface still clearly displays the "finger-like" outline. Within the cloud cavitation area, a re-entrant jet located at its tail can be observed. In subfigure (d), the re-entrant jet moves upstream along the "finger-like" vortex cavitation, inducing cavitation shedding towards the leading edge. A large-scale cloud cavitation appears at the trailing edge. In subfigure (e), the tail of the "finger-like" vortex cavitation fluctuates and sheds. The shedding cloud cavities collapse at the trailing edge, triggering the generation of the shockwaves. The shockwaves propagate upstream, causing the majority of the "finger-like" vortex

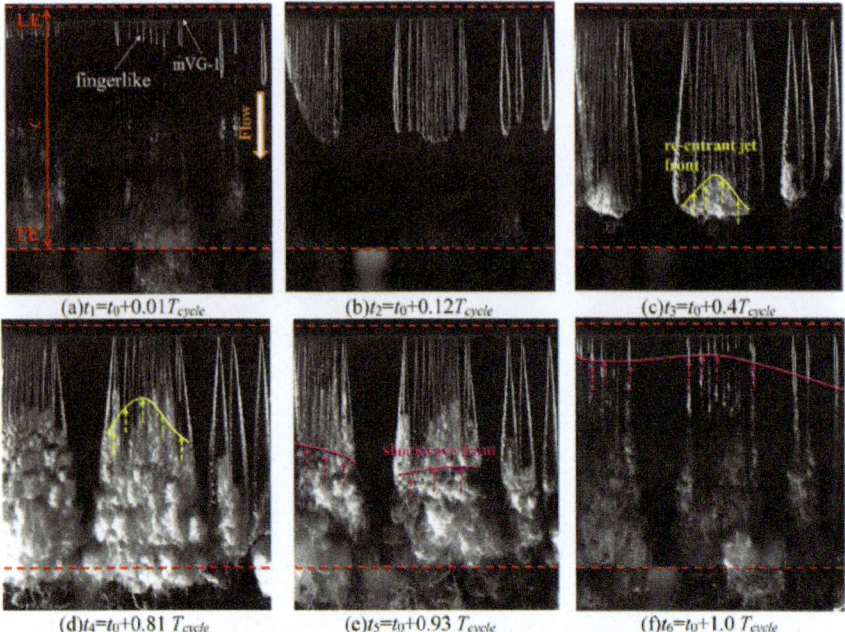

Fig. 3.20 High-speed photography of the cloud cavitation near the mVG-2 hydrofoil at 10,000 fps. Reprinted with the permission from Ref. [19] Copyright (2024) (ELSEVIER)

cavities to collapse. In subfigure (f), the leading edge maintains the "finger-like" vortex cavitation, which is non-uniform in the spanwise direction.

Figure 3.21 illustrates the photographs of the cloud cavitation near the symmetrical flat hydrofoils at 7,500 fps. In subfigure (a), the cloud cavitation area is notably small, with minimal temporal variation, suggesting a stable cavitation flow regime. As the cavitation number decreases, subfigure (b) depicts an enlargement of the cloud cavitation area, accompanied by an increase in complexity of the cavitation structure. As the cavitation number continues to decrease, subfigure (c) shows a more pronounced expansion of the cloud cavitation area, with a clear and distinct cloud cavitation region emerging at the cavitation inception stage. As depicted in subfigure (d), the cavitation number is the smallest, the cloud cavitation is marked by a significant escalation in both noise and vibration.

Figure 3.22 illustrates the photographs of the cloud cavitation near an axis-equipped hydrofoil at 7,000 fps. Initially, the cloud cavitation is observed above the rotational axis, progressively extending towards the trailing edge. Then, at 21.85 ms, the cavitation region reaches its fullest extent, with the cloud cavitation stretching to its maximum length. And, the cavitation region begins to retract towards the trailing edge. And, there is a concurrent transformation in the vortices. From 0 to 11.7 ms, the vortices near the rotational axis are predominantly clockwise, while those near the trailing edge exhibit an alternating pattern of the clockwise and counterclockwise

3.5 Cavitation Flow

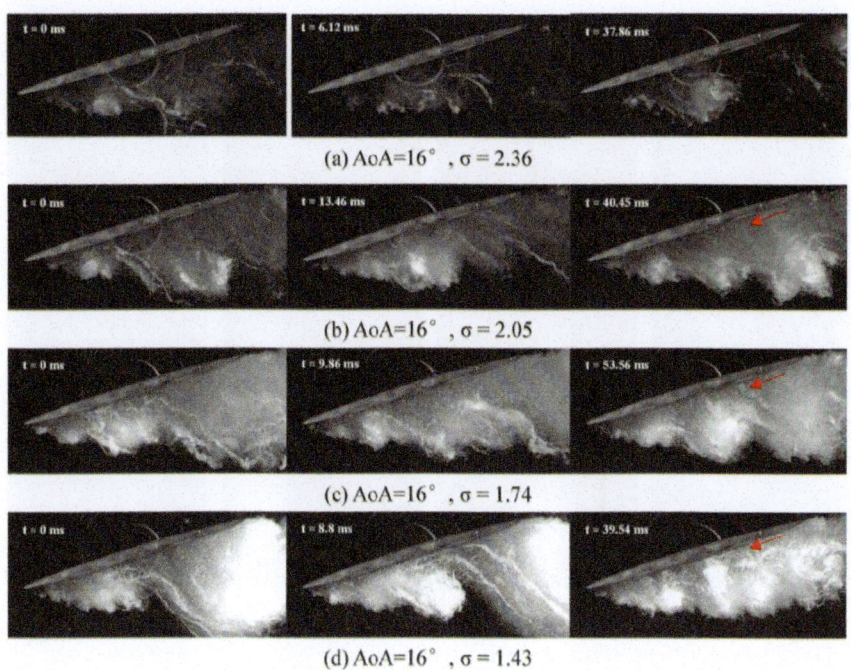

Fig. 3.21 High-speed photography of the cloud cavitation near the symmetrical flat hydrofoils at 7,500 fps. Reprinted with the permission from Ref. [20] Copyright (2023) (ELSEVIER)

rotations. At 14.95 ms, the vortices at the rotational axis diverge into two distinct groups, which rotate in opposite directions. The counterclockwise-rotating vortex converges with the cloud cavitation flow. At 19.95 ms, the pattern shifts to an intermittent rotation of the vortices, oscillating between clockwise and counterclockwise orientations.

3.5.3 Super Cavitation

During high-speed underwater navigation, objects may be trapped in vapor cavities, which is known as super cavitation. In this state, the majority of the object surface is sheathed by the vapor cavity, and the direct interaction between the fluid and the object only occurs at the leading edge.

Figure 3.23 illustrates the photographs of the super cavitation near a Tulin hydrofoil at 3,000 fps. The left subfigure shows the mixed super cavitation, with the cavity in a phase of dual-phase coexistence. Small-scale vortexes detach at the trailing edge, resembling cloud cavitation. The right subfigure shows the complete super cavitation, where a reduction in the cavitation number leads to the full development of the

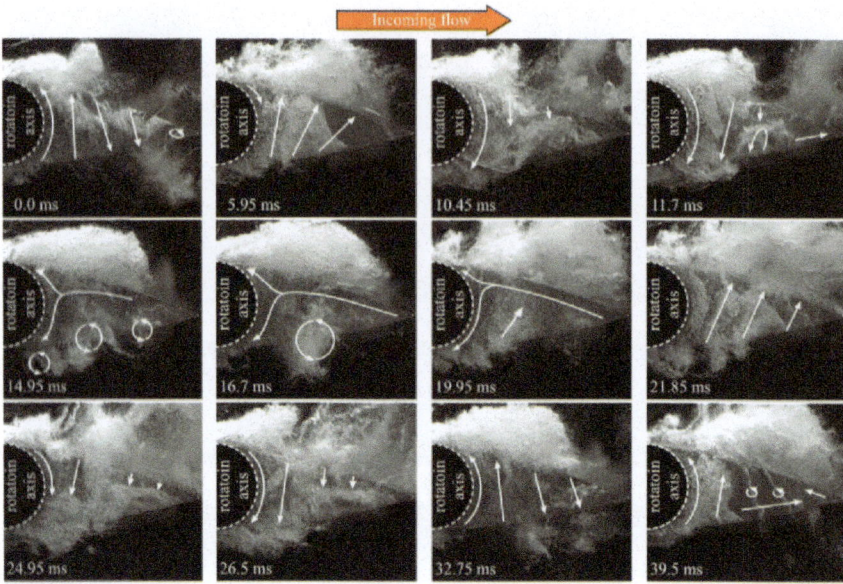

Fig. 3.22 High-speed photography of the cloud cavitation near an axis-equipped hydrofoil at 7,000 fps. Reprinted with the permission from Ref. [18] Copyright (2022) (ELSEVIER)

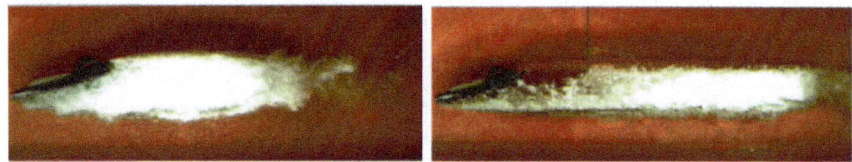

Fig. 3.23 High-speed photography of the super cavitation near the Tulin hydrofoil at 3,000 fps. Reprinted with the permission from Ref. [21] Copyright (2019) (ELSEVIER)

super cavitation. In the later stages of the super cavitation, the cavitation region is predominantly filled with vapor. A water–vapor mixing region is presented during the super cavitation evolution, characterized as a narrow band that oscillates within a defined range.

Figure 3.24 illustrates the photographs of the super cavitation near a blunt body at 3,000 fps. In the initial phase depicted on the left, the phenomena of fully developed super cavitation are observed. A well-defined cavity forms at the leading edge, and the cavity is punctured by the jet at the trailing edge. Subsequently, the jet advances towards the leading edge, with small bubble clusters peeling off. The subfigures on the right show a shorter cavity at the leading edge, which transitions into an elongated vapor–liquid mixed cavity and extends towards the trailing edge. Subsequently, the jet pushes towards the leading edge and causes a significant release of the bubbles that coalesce into a sizable cavity.

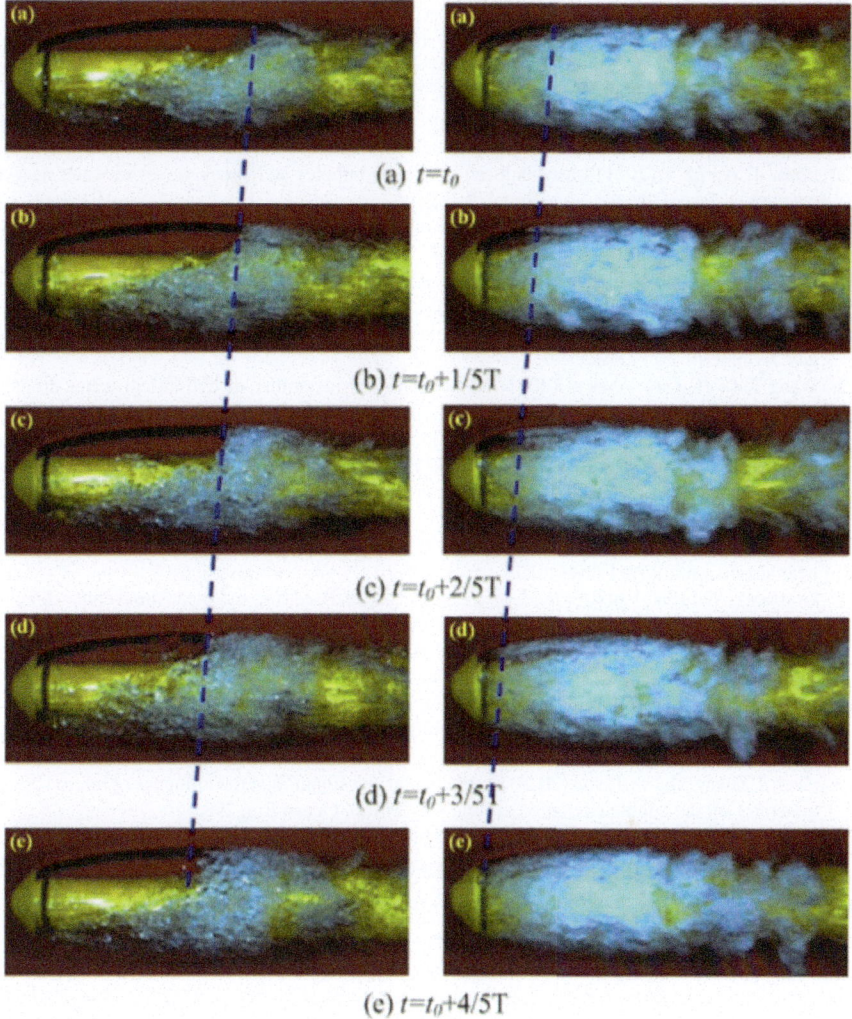

Fig. 3.24 High-speed photography of the super cavitation near a spherical blunt body at 3,000 fps. Reprinted with the permission from Ref. [22] Copyright (2017) (ELSEVIER)

References

1. Lauterborn W, Bolle H (1975) Experimental investigations of cavitation-bubble collapse in the neighbourhood of a solid boundary. J Fluid Mech 72(2):391–399
2. Brujan EA, Zhang AM, Liu YL et al (2022) Jet XE "Jet" ting and migration of a laser-induced cavitation bubble in a rectangular channel. J Fluid Mech 948:A6
3. Cui P, Zhang A, Wang S et al (2013) Experimental investigation of bubble dynamics near the bilge with a circular opening. Appl Ocean Res 41:65–75
4. Zhang J, Du Y, Liu J et al (2022) Experimental and numerical investigations of the collapse of a laser-induced cavitation bubble near a solid wall. J Hydrodyn 34(2):189–199

5. Vogel A, Lauterborn W, Timm R (1989) Optical and acoustic investigations of the dynamics of laser-produced cavitation bubbles near a solid boundary. J Fluid Mech 206:299–338
6. Brujan EA, Noda T, Ishigami A et al (2018) Dynamics of laser-induced cavitation bubbles near two perpendicular rigid walls. J Fluid Mech 841:28–49
7. Lindau O, Lauterborn W (2003) Cinematographic observation of the collapse and rebound of a laser-produced cavitation bubble near a wall. J Fluid Mech 479:327–348
8. Požar T, Agrež V (2021) Laser-induced cavitation bubbles and shock waves in water near a concave surface. Ultrason Sonochem 73:105456
9. Zwaan E, Le Gac S, Tsuji K et al (2007) Controlled cavitation in microfluidic systems. Phys Rev Lett 98(25):254501
10. Quah EW, Karri B, Ohl SW et al (2018) Expansion and collapse of an initially off-centered bubble within a narrow gap and the effect of a free surface. Int J Multiph Flow 99:62–72
11. Shen J, Ying J, Liu W et al (2024) The bubble dynamics near double cylinders within a narrow gap. Symmetry 16(7):841
12. Wang J, Li H, Guo W et al (2021) Rayleigh-Taylor instability of cylindrical water droplet induced by laser-produced cavitation bubble. J Fluid Mech 919:A42
13. Arora M, Ohl CD, Mørch KA (2004) Cavitation inception on microparticles: a self-propelled particle accelerator. Phys Rev Lett 92(17):174501
14. Ren Z, Zuo Z, Wu S et al (2022) Particulate projectiles driven by cavitation bubbles. Phys Rev Lett 128(4):044501
15. Teran LA, Rodriguez SA, Laín S et al (2018) Interaction of particles with a cavitation bubble near a solid wall. Phys Fluids 30(12)
16. Gonzalez-Avila SR, Van Blokland AC, Zeng Q et al (2020) Jetting and shear stress enhancement from cavitation bubbles collapsing in a narrow gap. J Fluid Mech 884:A23
17. Liu H, Guo Q, Shi L et al (2023) Lift-drag characteristics of S-shaped hydrofoil under different cloud cavitation conditions. Ocean Eng 278
18. Nichik MY, Timoshevskiy MV, Pervunin KS (2022) Effect of an end-clearance width on the gap cavitation structure: Experiments on a wall-bounded axis-equipped hydrofoil. Ocean Eng 254
19. Chen J, Zhang M, Liu T et al (2024) Experimental investigation of the influence of micro vortex generator on the bubble cavitation around a hydrofoil. Ocean Eng 298
20. Wang L, Tang F, Liu H et al (2023) Investigation on the vortex formation mechanism of flat hydrofoil under stall condition considering cavitation and non-cavitation states. Ocean Eng 285
21. Zhang M, Chen H, Wu Q et al (2019) Experimental and numerical investigation of cavitating vortical patterns around a Tulin hydrofoil. Ocean Eng 173:298–307
22. Liu T, Huang B, Wang G et al (2017) Experimental investigation of the flow pattern for ventilated partial cavitating flows with effect of Froude number and gas entrainment. Ocean Eng 129:343–351

Chapter 4
Visualization Research on Drop Dynamics

4.1 Introduction

In this chapter, the dynamic behaviors of the drop undergoing different movements are introduced. Firstly, the morphological variations are discussed when the drops impact onto different surfaces, which include thin liquid layers, wet surfaces, and dry surfaces. Then, the coalescence behaviors of the drops with other drops and free surfaces are presented. Finally, various types of the drop fragmentation mechanisms are explored, including drop splashing, pinch off, drop falling fragmentation, and metal drop fragmentation.

The impact and coalescence motions of the drops generally do not require very high shooting speeds (i.e., lower than 10,000 fps). When it comes to the microjet phenomena generated during the drop motion, due to the rapid generation, higher shooting speeds (i.e., higher than 100,000 fps) are required.

4.2 Drop Impact and Bouncing

4.2.1 Thin Liquid Layer

The drop impact is a complex process involving fluid dynamics, and its outcome is influenced by variables including liquid thickness, viscosity, and density.

Figure 4.1 illustrates the photographs of a water drop impacting onto the thin liquid layer with a low velocity at 6,000 fps. In the figure, a depression is observed at the position where the impact occurs, surrounded by a protrusion above the thin liquid layer. In the depression area, the solid surface is exposed to the air, and this area expands radially outwards together with the protrusion part. Subsequently, at 12.5 ms, the protrusion of the liquid layer reaches its maximum, with noticeable ripples

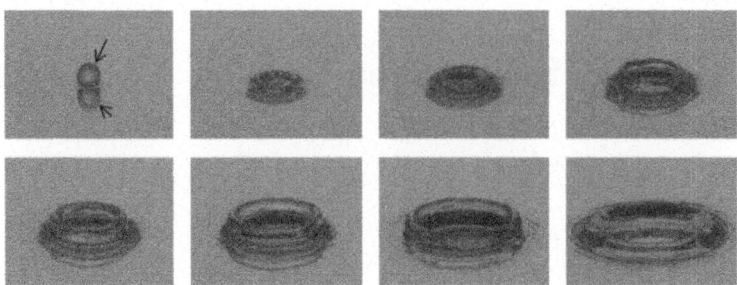

Fig. 4.1 High-speed photography of pure water drop impacting onto the thin liquid layer with a low velocity at 6,000 fps. Reprinted with the permission from Ref. [1] Copyright (2021) (ELSEVIER)

forming around it. The energy propagates outward as waves, and the protrusion liquid falls with the flow back into the depression area.

As the drop impact velocity increases, the drop-liquid interaction exhibits more complex phenomena. Figure 4.2 illustrates the photographs of a drop impacting onto the thin liquid layer with a high velocity at 6,000 fps. As shown in the figure, when the drop impacts onto a thin liquid layer, a rim of the water film is observed, centered at the point of the impact. And, a cluster of small drops forms atop the liquid layer. Then, the growth pattern of the layer is observed to transition from an inverted conical shape to a cylindrical shape, and finally to a spherical growth pattern. The exposed area of the solid surface at the center expands in a circular shape. The cluster of the small drops is formed during the initial impact and coalesces with subsequently ejected drops, resulting in the formation of larger drops. This process is clearly observable at 12.8 ms. Additionally, ripples form on top of the liquid layer. At 4 ms, the ripples are small, and gradually grow and persist at the surface. At 15.8 ms, the ripples rapidly propagate towards the base of the layer. By 29.3 ms, the ripples have completely disrupted the layer. After 33.7 ms, the small drops fall back and impact onto thin liquid layer again, disrupting the motion of the liquid surface in the impact area.

Figure 4.3 illustrates the photographs of a drop (55% glycerol) impacting onto the thin liquid layer with a high velocity at 6,000 fps. As shown in the figure, after the drop impacts onto the thin liquid layer, a rim of water film is formed around the impact area, with the solid surface exposed to the air. Then, the water film develops upwards and reaches its highest point at 7.83 ms. At 18.3 ms, the water film begins to fall and transforms into a ring-like protrusion on the liquid surface. Concurrently, with the formation of the water film, numerous smaller drops are produced at the top of the film. These drops gradually coalesce into larger drops as the water film extends upwards. When the water film falls, its descent speed is higher than that of the drops, causing the drops to separate from the water film, which can be observed at 18.3 ms. Furthermore, waves are formed around the ring-like protrusion, and propagate outwards.

Figure 4.4 illustrates the photographs of the drop (ethanol) impacting onto the thicker liquid layer with a high velocity at 6,000 fps. As shown in the figure, after the drop impacts onto the liquid layer, an inverted conical water film is initially formed.

4.2 Drop Impact and Bouncing 39

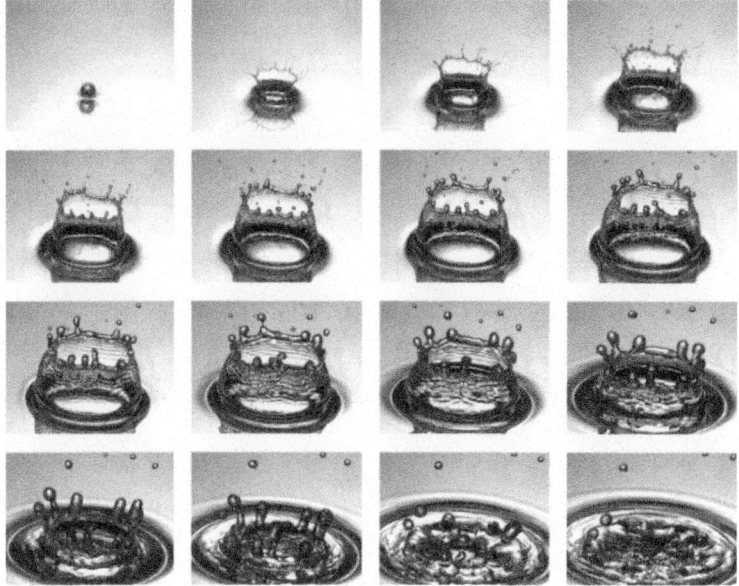

Fig. 4.2 High-speed photography of pure water drop impacting onto the thin liquid layer with higher velocity at 6,000 fps. Reprinted with the permission from Ref. [1] Copyright (2021) (ELSEVIER)

Fig. 4.3 High-speed photography of a glycerol water mixture drop impacting onto the thin liquid layer with a high velocity at 6,000 fps. Reprinted with the permission from Ref. [1] Copyright (2021) (ELSEVIER)

By 2.67 ms, the water film transitions to a cylindrical expansion. And by 4.33 ms, it further extends into a spherical shape. At 12.7 ms, the top of the spherical cap aggregates. At the onset of the water film formation, a multitude of small drops form at the top, which gradually coalesce into larger drops as the film evolves. And by 12.7 ms, they collide to form a liquid column. Concurrently, the impact-induced circular depression expands over time, and no wave formation is observed.

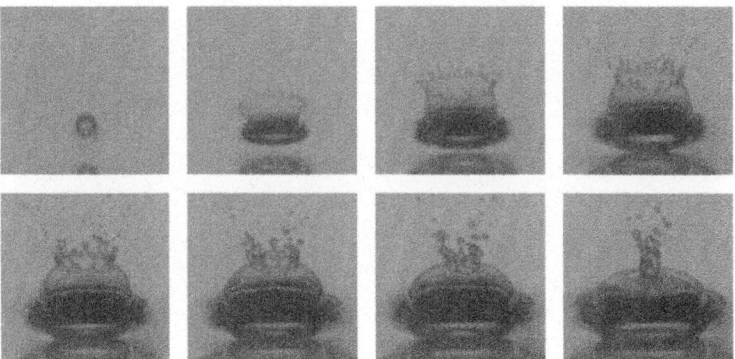

Fig. 4.4 High-speed photography of the ethanol drop impacting onto the thicker liquid layer with higher velocity at 6,000 fps. Reprinted with the permission from Ref. [1] Copyright (2021) (ELSEVIER)

4.2.2 Wetted Surface

When a drop impacts onto the wetted surface, two primary phenomena occur: the drop rebounds and the drop spreads. And, these phenomena are closely related to the drop density, viscosity, and impact velocity.

Figure 4.5 illustrates the photographs of a drop (butanol) impacting onto the wetted cylindrical surface (with its centerline perpendicular to the view) at 10,000 fps. The radius of the cylinder is much larger than that of the drop. When impacting with the wetted surface, the drop exhibits like a bouncing ball hitting a solid plane. Specifically, the bottom of the drop first expands outward in a cylindrical shape, and the top retains spherical. At 4.8 ms, the top of the drop becomes flat, taking on a pancake-like shape. Subsequently, the sides of the drop begin to contract while the top bulges. By 9.5 ms, the drop takes on an egg-like shape. After 9.5 ms, the drop rebounds from the wetted surface. During the process of the bouncing and falling back, the drop undergoes multiple shape variations. At 14.8 ms, the drop adopts an ellipsoidal shape, with its major axis oriented horizontally. At 23.2 ms, it transforms into an ellipsoid with its major axis oriented vertically. By 27.9 ms, the drop once again presents an ellipsoidal shape with its major axis oriented horizontally. At this point, it can be observed that the center of the drop remains virtually stationary. Eventually, the drop falls in a vertically oriented ellipsoidal shape and makes contact with the surface in a spherical shape at 40.4 ms.

When the drops impact surfaces with varying curvatures at similar velocities, they exhibit different dynamic behaviors. Figure 4.6 illustrates the photographs of a drop (butanol) impacting onto the wetted cylindrical surfaces with different curvatures at 10,000 fps. Subfigures (a)–(d) demonstrate the varying phenomena as the curvature of the cylinder decreases. In subfigure (a), the cylinder has a largest curvature, the liquid flows laterally upon impact, forming a curved liquid layer. In subfigure (b), with a decreased curvature, the liquid layer expands, and an outward depression

4.2 Drop Impact and Bouncing

Fig. 4.5 High-speed photography of drop impacting onto the wetted surface with low velocity at 10,000 fps. Reprinted with the permission from Ref. [2] Copyright (2013) (ELSEVIER)

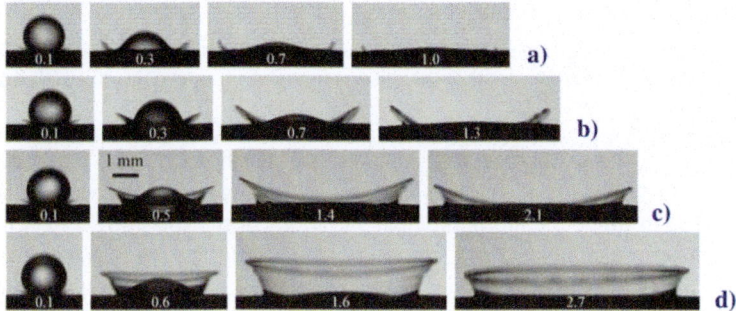

Fig. 4.6 High-speed photography of drop impacting onto the wetted surfaces with different curvatures at 10,000 fps. Reprinted with the permission from Ref. [3] Copyright (2014) (ELSEVIER)

of the film is observed. In subfigure (c), the cylinder curvature is further reduced, and the liquid layers form not only laterally but also anteriorly and posteriorly. The impact forms a toroidal liquid layer with a top contour resembling a saddle surface. In subfigure (d), the curvature is minimal and the surface can be considered flat, the liquid layer maintains a consistent horizontal level at the top.

4.2.3 Dry Surface

The drop impact phenomena onto dry surfaces exhibit significant differences from those observed on wetted surfaces or thin liquid layers. On dry surfaces, there is no other liquids to guide and lubricate the impacting drop, which means that the part of the drop in contact with the plane encounters greater resistance.

Figure 4.7 illustrates the photographs of different drops impacting onto the dry surface with different velocities at 10,000 fps. In subfigure (a), a silicone oil drop impacts onto the dry surface at 1.4 m/s, forming a liquid layer at the center, which spreads outwards. During the impact, the drop is squeezed, continuously supplying liquid to the film, creating a pancake-like liquid layer. In subfigure (b), a glycerin-water mixture drop impacts onto the dry surface at 2.8 m/s, forming a circular liquid layer. Tiny drops are observed to detach from the front edge of the film, which is known as immediate splashing. Subfigure (c) shows a silicone oil drop impacting onto the dry surface at 2.5 m/s. After the impact, in addition to the formation of a

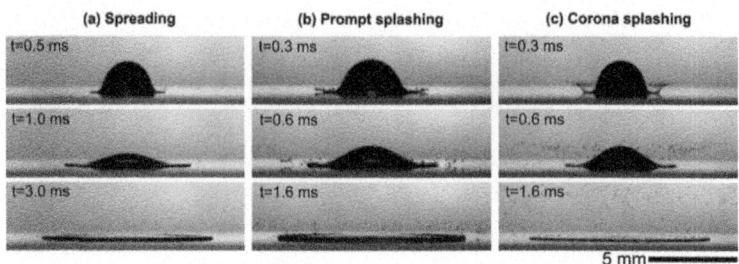

Fig. 4.7 High-speed photography of different drops impacting onto the dry surface with different velocities at 10,000 fps. Reprinted with the permission from Ref. [4] Copyright (2019) (ELSEVIER)

spreading liquid layer, the film also lifts upward and breaks. And, a multitude of small drops is produced and splashes outwards, which is known as crown splashing.

To further examine the morphological variations of the splashing, Fig. 4.8 illustrates the photographs of various drops impacting onto the dry surface with high velocities at 100,000 fps. As shown in subfigure (a), a wine drop impacts onto the dry surface at a high velocity, and exhibits a typical crown splashing. Specifically, at the initial stage, a toroidal protrusion forms immediately, which then rapidly spreads outwards. Then, the toroidal protrusion lifts upward, and evolves into a toroidal liquid layer. Concurrently, the liquid layer breaks apart after a certain period, generating a multitude of tiny drops. Over time, the liquid layer on the dry surface spreads outward, with its leading edge persistently lifting and breaking. And, a continuous stream of small drops is formed. In subfigure (b), a water drop impacts onto the dry surface at a high velocity, and demonstrates a classic immediate splashing. Upon contact with the dry surface, the drop immediately spreads into a liquid layer on the surface. Then, the leading edge of the film breaks apart almost instantaneously, producing a plethora of the minuscule drops. Notably, unlike the crown splashing, the leading edge in an immediate splashing does not lift up significantly. Instead, more liquid accumulates only in the front region.

The impact of the surface roughness is also significant. Figure 4.9 illustrates the photographs of the drop impacting onto the wooden surfaces with varying roughness at 75,000 fps. As shown in the figure, drops are released from a height of 40 cm. When the surface is relatively smooth (with a roughness of 0.51 μm), a liquid layer forms and spreads outward. A toroidal protrusion at the leading edge of the liquid layer spreads outward, and small drops detach. As the surface roughness increases, the formation time of the toroidal protrusion at the leading edge is delayed. Additionally, the distribution of the drops detaching from the leading edge becomes highly irregular. When the surface roughness reaches 2.03 μm, the toroidal protrusion at the leading edge disappears, replaced by an inhomogeneous distribution of the drops spreading outward. With further increases in roughness, the liquid layer area decreases, and the liquid is ejected outward from the film in the form of small drops. Overall, as surface roughness increases, the liquid layer encounters greater

4.2 Drop Impact and Bouncing

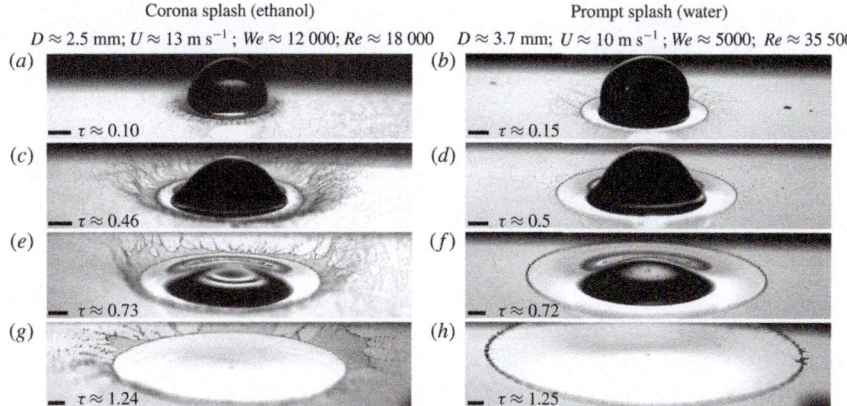

Fig. 4.8 High-speed photography of various drops impacting onto the dry surface with high velocities at 100,000 fps. Reprinted with the permission from Ref. [5] Open access under a CC BY 4.0 license, https://creativecommons.org/licenses/by/4.0/

resistance during its outward spreading. This resistance hinders the expansion of the liquid layer.

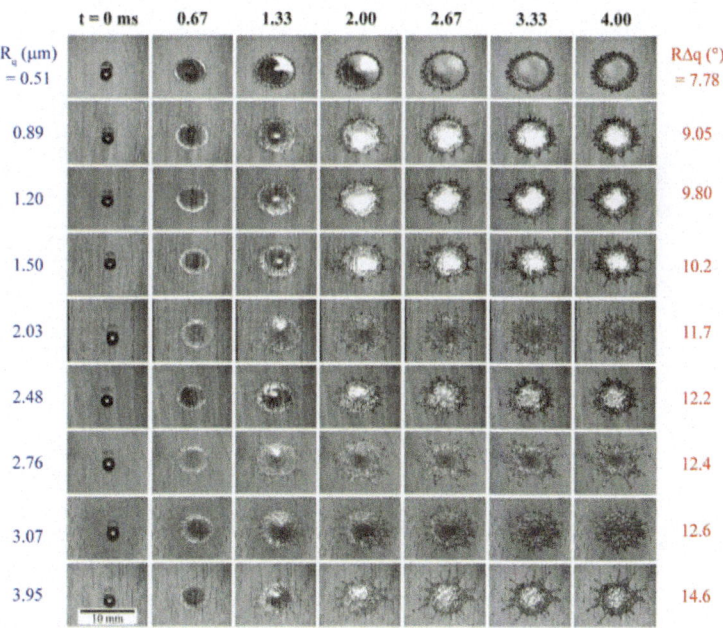

Fig. 4.9 High-speed photography of drops impacting onto the rough-dry surface at 75,000 fps. Reprinted with the permission from Ref. [6] Copyright (2021) (ELSEVIER)

4.3 Drop Coalescence

4.3.1 Drop-Flat Interface

The process of the drop coalescence with a free surface is influenced by the liquid density, viscosity, and solubility.

Figure 4.10 illustrates the photographs of the hydrolyzed polyacrylamide drops coalescing with silicone oil in an electric field at 5,000 fps. In subfigure (a), with an electric field strength of 1200 V/cm, the drop achieves complete coalescence with the free surface. The drop first makes contact with the free liquid surface, forming a neck-like channel. Then, the liquid within the drop flows towards the free surface. As the liquid at the top of the drop diminishes, the neck channel expands into a conical channel. Eventually, the drop forms a conical contraction. In subfigure (b), the electric field strength is increased to 2200 V/cm, the partial coalescence occurs between the drop and the free surface. A neck-like channel also forms, allowing the liquid within the drop to flow into the external fluid. Then, the neck channel does not expand into a conical channel but instead becomes thinner, which is associated with the increased electric field strength. Ultimately, the neck channel closes, and the drop hovers above the free surface. In subfigure (c), at an electric field strength of 2600 V/cm, no coalescence occurs. Specifically, the drop makes contact with the free surface under the influence of the gravity, then separates from it. The gravitational acceleration of the drop is not significant enough for the liquid surface tension to counteract the electric field force. It is similar to the scenario where small drops detach from the free surface after partial coalescence. In addition, the surface slightly protrudes upwards, forming a conical structure. Once the drop detaches, the conical structure gradually dissipates.

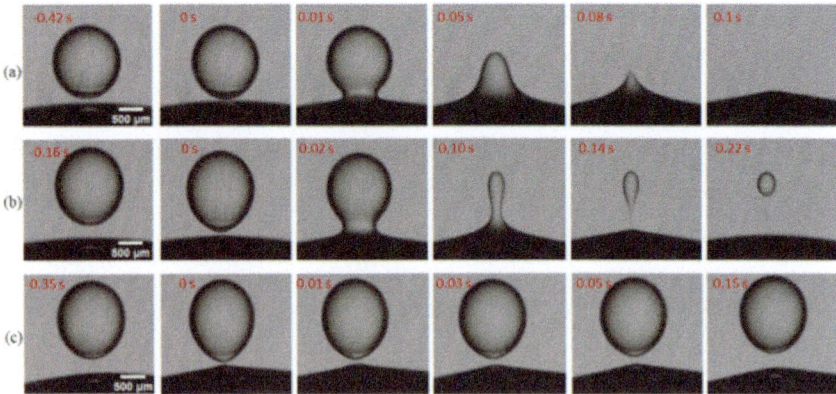

Fig. 4.10 High-speed photography of a drop coalescing with the liquid surface at 5,000 fps. Reprinted with the permission from Ref. [7] Copyright (2024) (ELSEVIER)

4.3 Drop Coalescence

Figure 4.11 illustrates the photographs of a diluted ink drop coalescing with a free surface at 4,000 fps. As shown in the figure, upon contact with the free surface, a channel is formed, and the liquid within the drop flows into the external fluid through it. Over time, the channel transitions from a cylindrical shape (at 1.25 ms) to a conical shape (at 6.5 ms). The top of the drop then tapers to a sharp point while the channel maintains its shape relatively well (at 8.25 ms). Subsequently, the channel is compressed into a conical shape. After the drop fully penetrates the free liquid surface, a downward indentation appears due to gravity. Around this indentation, a toroidal vortex forms (at 14.5 ms). As the indentation expands, the vortex becomes more pronounced. At 28.0 ms, the indentation contracts, and the toroidal vortex fully forms, moving downward.

Figure 4.12 illustrates the photographs of the more diluted ink drop coalescing with a free surface at 4,000 fps. The liquid within the drop flows out, causing the drop to taper to a conical shape. At 11.00 ms, the drop is completely submerged within the free surface. Subsequently, the kinetic energy carried by the drop creates a downward-facing cavity. This cavity initially expands in a spherical manner. As the cavity grows to a certain size, a conical protrusion forms at the bottom of the cavity, representing a smaller cavity. At 20.25 ms, the growth of the smaller cavity ceases. The conical protrusion starts to contract at its top, flattening the base of the recessed cavity. Subsequently, the recessed cavity contracts in a frustum shape.

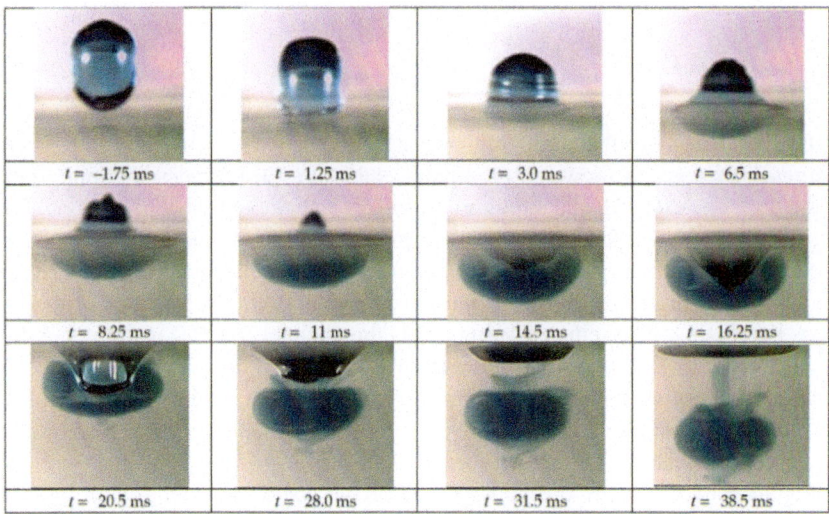

Fig. 4.11 High-speed photography of a diluted ink drop coalescing with a free surface at 4,000 fps. Reprinted with the permission from Ref. [8] Open access under a CC BY 4.0 license, https://creativecommons.org/licenses/by/4.0/

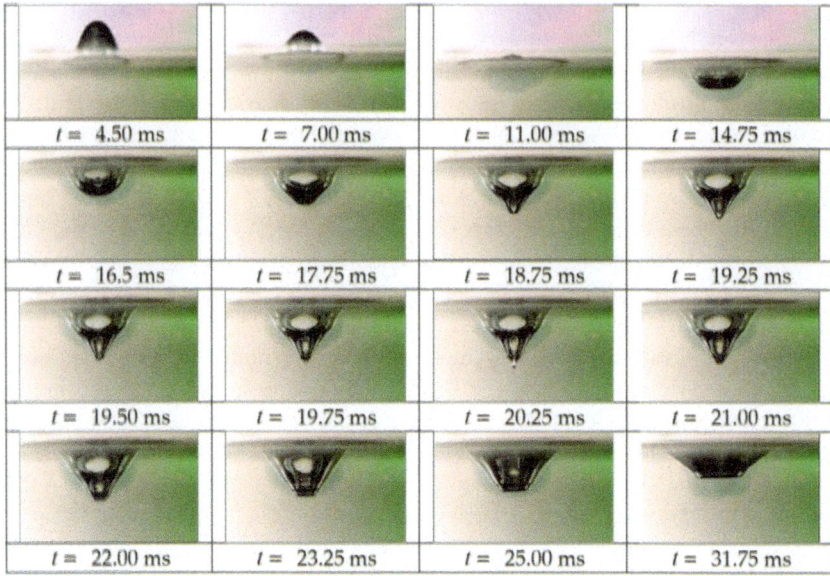

Fig. 4.12 High-speed photography of a more diluted aqueous ink drop coalescing with a free surface at 4,000 fps. Reprinted with the permission from Ref. [8] Open access under a CC BY 4.0 license, https://creativecommons.org/licenses/by/4.0/

4.3.2 Drop-Drop

Figure 4.13 illustrates the photographs of the coalescence of two drops with similar radii at 5,000 fps. As shown in the figures, subfigure (a) provides a primary view of the drop coalescence, and subfigure (b) offers an enlarged view of subfigure (a). In subfigure (a), a cylindrical channel forms at the surface where the drops meet. The fluids within both drops flow through this cylindrical channel. Subsequently, the cylindrical channel expands into a neck-like channel, with the volumes of the two drops decreasing. The reduced volume fills the channel. At 800 μs, the neck-like channel becomes quite pronounced. Afterward, the neck-like channel continues to expand, and the channel body reverts to a cylindrical shape, resulting in a dumbbell-shaped coalescence of the two drops. In subfigure (b), it is noted that the surface of the drops connected by the channel becomes flattened.

Figure 4.14 illustrates the photographs of the coalescence of two drops with different radii at 5,000 fps. As shown in the figures, the drops form a cylindrical channel. Then, the cylindrical channel expands, transforming into a neck-like channel, with the neck moving upward. The surface of the smaller drop becomes flattened, while the surface of the larger drop does not exhibit significant variations.

Figure 4.15 illustrates the photographs of a water drop within an oil drop merging with an external drop at 10,000 fps. As shown in the figure, the drop descends and makes contact with the oil–water interface, resulting in the formation of a neck-like

4.3 Drop Coalescence

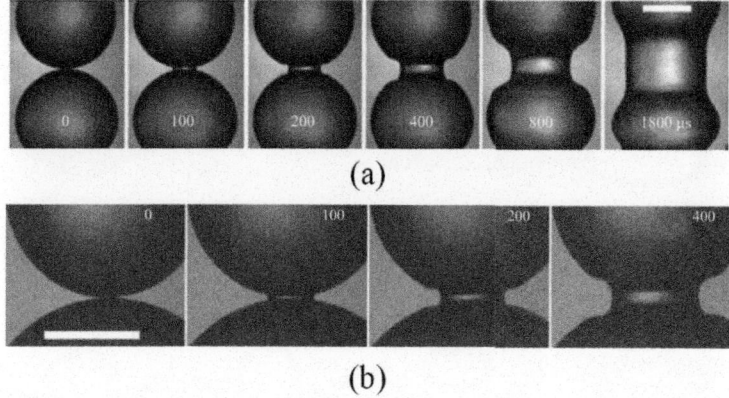

Fig. 4.13 High-speed photography of the coalescence of two drops at 5,000 fps. Reprinted with the permission from Ref. [9] Copyright (2005) (Cambridge University Press)

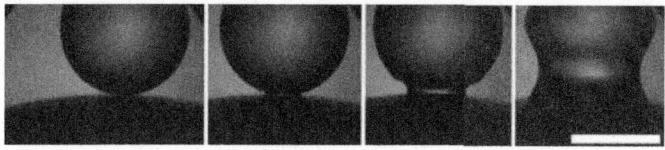

Fig. 4.14 High-speed photography of the coalescence of two drops with different radius at 5,000 fps. Reprinted with the permission from Ref. [9] Copyright (2005) (Cambridge University Press)

channel. Subsequently, the top of the water drop is compressed, causing the liquid within the drop to flow into the external environment through the neck-like channel. And, the drop volume reduces, with the neck of the channel becoming smoother. At 3 ms, the neck-like channel disappears, and the drop takes on an elongated, vertically oriented elliptical shape. Then, the top and bottom of the drop contract and the drop expands horizontally, forming an egg-like shape. Subsequently, the drop continues to deform, adopting an elongated, horizontally oriented elliptical shape. At 6 ms, the drop exhibits a spherical shape and comes into contact with the oil–water interface, but no channel is formed.

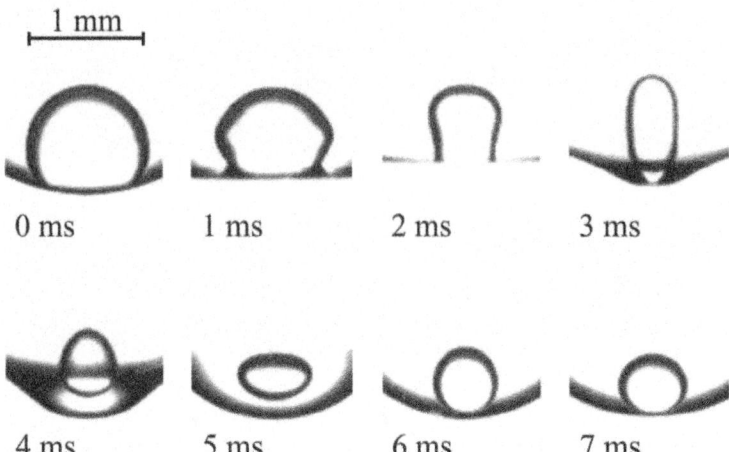

Fig. 4.15 High-speed photography of coalescence of water drop within an oil drop merging with an external drop at 10,000 fps. Reprinted with the permission from Ref. [10] Copyright (2011) (ELSEVIER)

4.4 Drop Fragmentation

4.4.1 Drop Splashing

When a drop is subjected to a strong internal or external force, it can no longer maintain its current shape, leading to the surface rupture and the production of the smaller drops and jets. This complex variation in drop morphology is known as the splashing.

Figure 4.16 illustrates the photographs of the drop splashing in the center at 40,000 fps. As shown in the figure, a laser-induced bubble is focused inside the drop, which generates a significant amount of energy upon collapse. And, it greatly changes the internal flow field environment of the drop and causes the surface to rupture and deviate from a spherical shape. Specifically, at 350 μs, the bubble completely collapses and begins to expand again, causing the surface to break apart into smaller drops. Subsequently, the splashing intensifies, and further disrupts the main body of the drop until it can no longer maintain a spherical shape (2225 μs). It is noteworthy that the smaller drops formed from the ruptured surface continue to coalesce into larger drops as they splash outward. Concurrently, during the splashing process, a liquid film forms at the part connected to the drop surface, which expands and lifts upward over time. It is not until 1275 μs that the raised liquid film breaks.

Figure 4.17 illustrates the photographs of the off-center drop splashing at 40,000 fps. As shown in the figure, after the bubble collapses, it first destroys the stability of the right drop surface. The breakage of the right surface generates a large number of small drops and shoots out to the right, while the left surface still maintains its

4.4 Drop Fragmentation

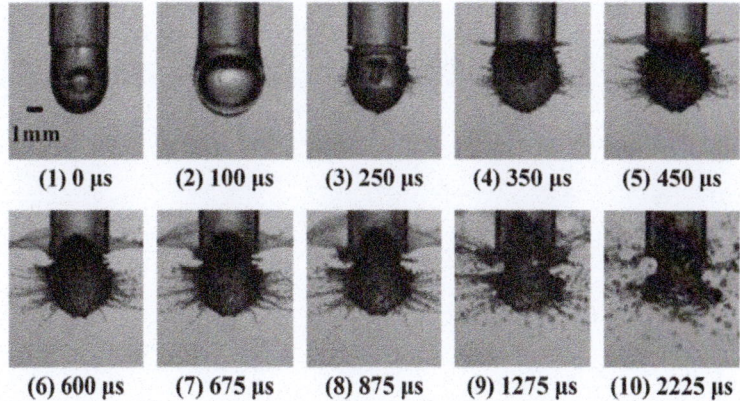

Fig. 4.16 High-speed photography of the drop splashing in the center at 40,000 fps. Reprinted with the permission from Ref. [11] Open access under a CC BY 4.0 license, https://creativecommons.org/licenses/by/4.0/

previous shape. As the influence of the bubble collapse spreads to the left side, at 1025 μs, a protrusion appears on the left surface. Subsequently, the left protrusion shoots out to the left in the form of a liquid column. The broken surface on the left does not form a large number of small drops.

Figure 4.18 illustrates the photographs of the free-falling drop splashing at 1,000,000 fps. In subfigure (a), the drop radius is 186 μm and the laser energy is 4.9 mJ. The drop is relatively small and there is slight atomization during the falling process. As the bubble collapses, the atomization of the drop intensifies and is completely atomized after 25.0 μs. In subfigure (b), the drop radius is 401 μm and the laser energy is 2.7 mJ. The drop is relatively large and there is no atomization phenomenon during the falling process. As the bubble collapses, a large number of small drops are formed on the right side and a liquid film is formed on the left side.

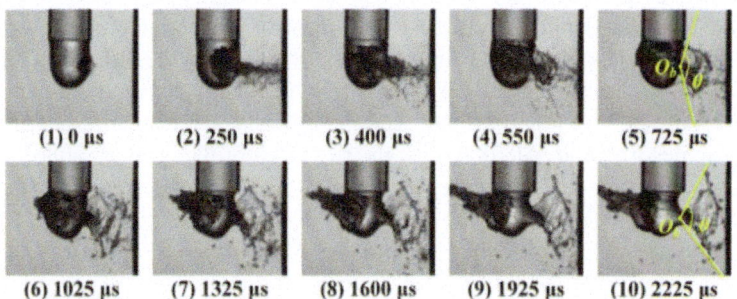

Fig. 4.17 High-speed photography of the off-center drop splashing at 200,000 fps. Reprinted with the permission from Ref. [12] Open access under a CC BY 4.0 license, https://creativecommons.org/licenses/by/4.0/

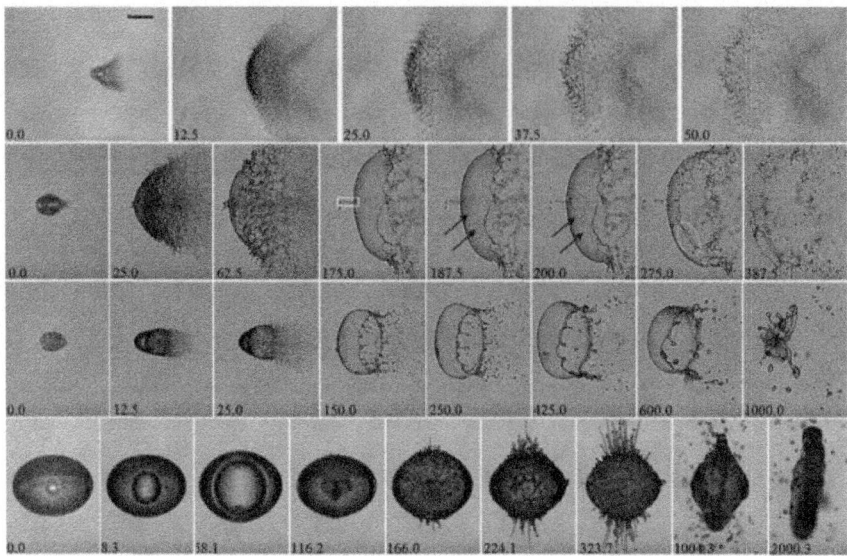

Fig. 4.18 High-speed photography of the free-falling drop splashing at 1,000,000 fps. Reprinted with the permission from Ref. [13] Copyright (2016) (Cambridge University Press)

Different from subfigure (a), the small drops formed by the breakup of the drop do not show an atomization phenomenon. In subfigure (c), the drop radius is 450 μm. The right surface of the drop is destroyed by the bubble collapse and small drops are formed. Similar to subfigure (a), an atomization phenomenon occurs. Subsequently, the surface on the right side enters the air and the entire drop is stretched into a liquid film. Different from subfigure (b), the liquid film in subfigure (c) is more complete. During the falling process, the liquid film is first stretched and then the range of the liquid film becomes smaller. And, the drops on the right edge of the liquid film become larger. In subfigure (d), the drop radius is 1419 μm and the laser energy is 2.2 mJ. The drop is relatively large, so that the bubble grows completely inside the drop. At 166.0 μs, the bubble collapses and the drop surface stability is destroyed. Small drops and jets splash out from the drop surface, and at the same time, irregular depressions and protrusions can be observed. Subsequently, the entire drop becomes unstable. At 323.7 μs, liquid columns splash outwards, and the entire surface of the drop becomes irregular. At 1004.3 μs, the two sides of the drop are squeezed, and the upper and lower sides protrude outwards.

4.4.2 Drop Pinch-Off

When a drop is subjected to gravity or other external forces, its shape changes. If the surface tension is difficult to maintain the continuous deformation of the drop, the drop breaks from the weak point, which is known as the drop pinching off.

Figure 4.19 illustrates the photographs of the drop pinch off at 2,000,000 fps. As shown in the figure, under the influence of the gravity, the drop moves downward slowly, and a neck-like structure appears at the junction of the drop and the nozzle liquid. Then, the neck-like part in contact with the drop begins to become smaller and present a conical structure. Subsequently, the drop detaches and falls, and the area around the neck becomes flat. After the drop detaches, the neck-like structure rebounds upward, but it does not merge with the liquid at the nozzle. Instead, after rebounding, the neck-like part in contact with the nozzle liquid becomes smaller, takes on a conical structure and falls off. And, it descends in a rotational symmetric structure. Similarly, the position where the nozzle liquid contacts the neck-like part also becomes flat.

Figure 4.20 illustrates the photographs of a microjet pinching off into drops at 14,000,000 fps. As shown in the figure, the nozzle generates a microjet with a radius of 1.25 μm. During the falling process of the jet, multiple neck-like structures appear in the main body. The neck-like structure at the end of the jet breaks off to form a conical drop, and then the drop evolves into a spherical drop. Simultaneously, the next-level neck-like structure at the end of the jet breaks off to form a drop with a horizontal major axis. Then, the drop also evolves into a spherical drop. Sometimes the neck-like structure at the end does not break off first, but the next-level neck-like structure breaks off to form a dumbbell-shaped drop. Subsequently, the middle of the dumbbell-shaped drop breaks off to form two drops that fall in sequence. When the jet is longer, the rate of which the jet pinches off into drops becomes faster, and it allows the shorter jet to continuously grow. In general, after the microjet becomes longer and drops fall, the jet length decreases. Then, the dripping speed of the shorter jet becomes slower and the jet becomes longer, repeating in a cycle.

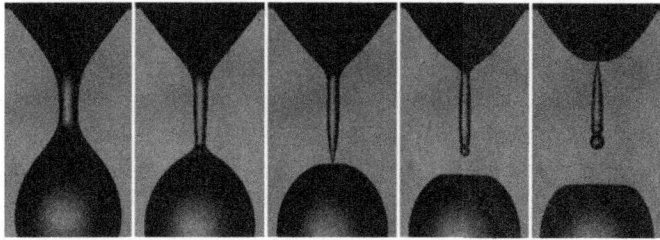

Fig. 4.19 High-speed photography of the drop pinch-off at 2,000,000 fps. Reprinted with the permission from Ref. [14] Copyright (2007) (AIP Publishing)

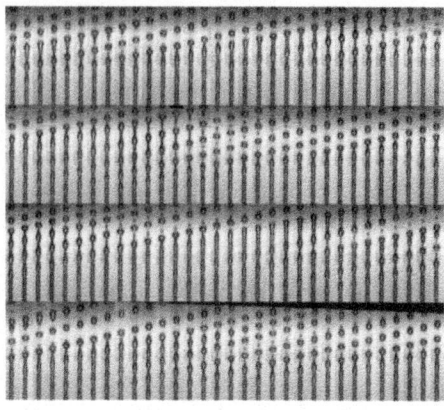

Fig. 4.20 High-speed photography of the drop pinch-off at 14,000,000 fps. Reprinted with the permission from Ref. [15] Copyright (2010) (AIP Publishing)

4.4.3 Drop Falling Fragmentation

During the falling process, the drop shatters under the influence of an airflow coming from the opposite direction. When the drops are small enough, this phenomenon can also occur during the free-falling process.

Figure 4.21 illustrates the photographs of the free-falling drop at 1,600 fps. As shown in the figure, the diameter of the drop is 2.9 mm. In subfigure (a), the incoming air velocity is relatively small, and the drop remains spherical. In subfigure (b), as the incoming air velocity increases, the incoming side of the drop becomes flat, and the opposite side becomes ellipsoidal. At this point, the drop still maintains an intact shape during the falling process. In subfigure (c), the incoming flow velocity further increases, and the incoming side of the drop indents into the it. Subsequently, a cavity appears inside the drop, and the side opposite to the incoming flow turns into a liquid film. After that, the liquid film breaks and the drop becomes annular. Then, the annular drop is blown apart and divided into drops on the left and right sides, accompanied by small drops formed by the surface fragmentation. In subfigure (d), the incoming side of the drop is indented, and a liquid film also appears on the other side. Compared with subfigure (c), the liquid film is larger and the annular area is smaller. In addition, the annular drop does not break into two drops but is blown apart by the incoming flow.

Figure 4.22 illustrates the photographs of the free-falling drop in airflow with different temperatures at 10,000 fps. In subfigure (a), at the room temperature, drops are affected by different mixed gases with the same velocity. Specifically, when the incoming gas density is 1.025 kg/m^2, the drop changes from a spherical shape to the flat shape. Subsequently, the incoming side is blown into a bag-shaped liquid film with a circular edge. After that, the liquid film breaks and generates a large number of small drops. As the incoming flow density decreases, the time for the drop to turn into a bag-shaped liquid film is delayed, and there are relatively intact drops after the liquid film breaks. In particular, when the incoming flow density is 0.716 kg/m^2, a bag-shaped liquid film is not formed. Instead, the flat drop breaks in the middle to

4.4 Drop Fragmentation

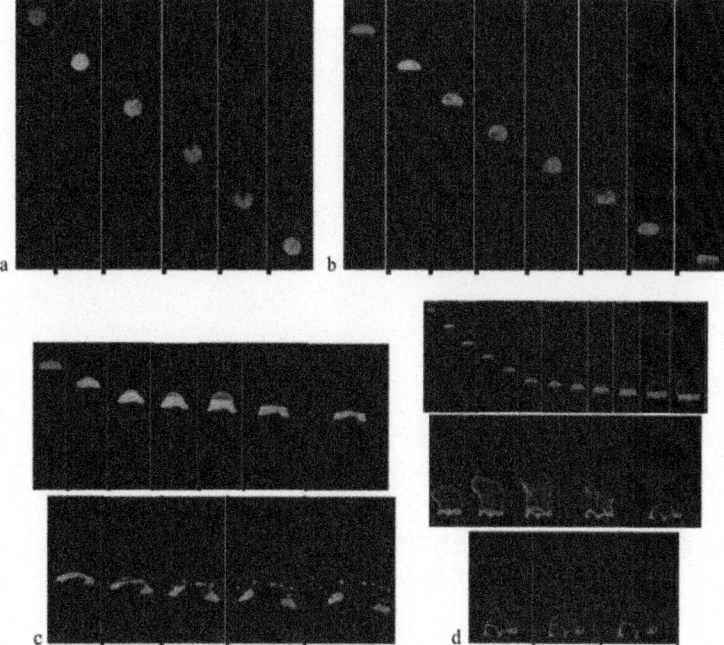

Fig. 4.21 High-speed photography of the free-falling drop at 1,600 fps. Reprinted with the permission from Ref. [16] Copyright (2023) (ELSEVIER)

form two smaller drops. In subfigure (b), the incoming flow temperature increases and the density decreases. When the temperature is 413 K, the drop transforms into a distorted flat structure, which is different from a lower temperature. As the incoming flow temperature increases and the density decreases, the appearance time for the bag-shaped liquid film structure is advanced. In general, an increase in temperature accelerates the break of the drop.

4.4.4 Metal Melt Drop Fragmentation

When a metal melt drop falls into the liquid, the liquid drop cools rapidly. The huge temperature difference between the interior and exterior surfaces of the drop causes the metal melt drop to break, and smaller drops are generated. Figure 4.23 illustrates the photographs of the fragmentation of a metal melt drop dripping into the liquid at 6,000 fps. As shown in the figure, after the metal melt drop contacts the liquid, the bottom of the metal melt drop breaks firstly, and a needle-like metal column protrudes outward. As the volume of the metal melt drop entering the liquid increases, the entire metal melt drop shows outward needle-like protrusions, and the liquid around the metal melt drop is evaporated to form a large number of the bubbles. During the

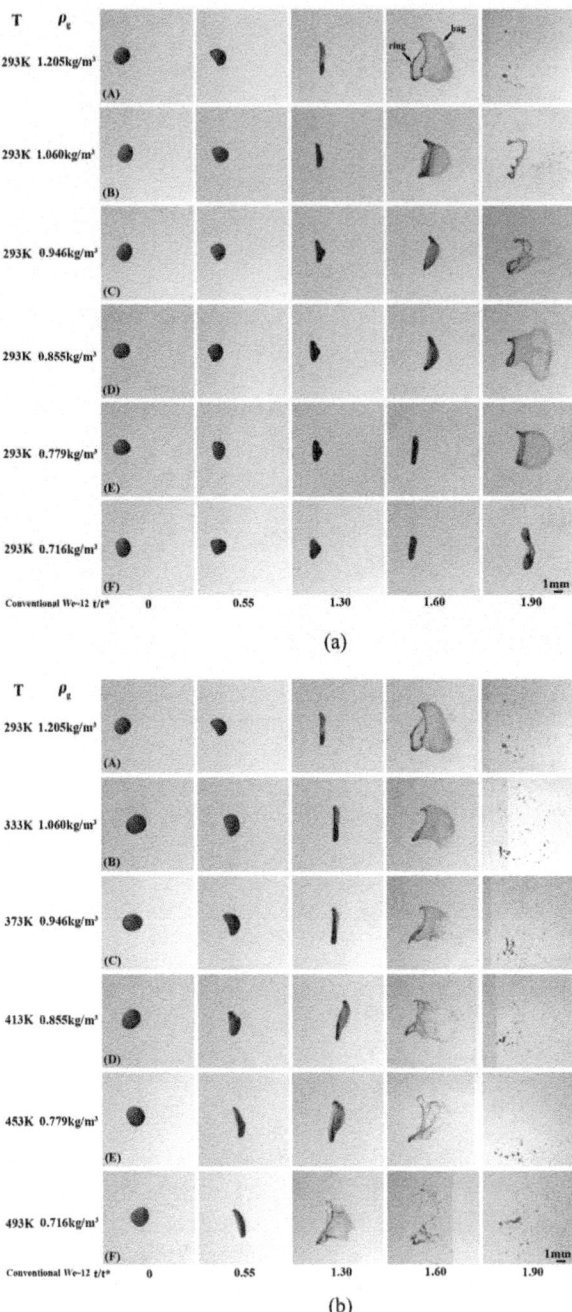

Fig. 4.22 High-speed photography of free-falling drop in airflow with different temperatures at 10,000 fps. Reprinted with the permission from Ref. [17] Copyright (2024) (AIP Publishing)

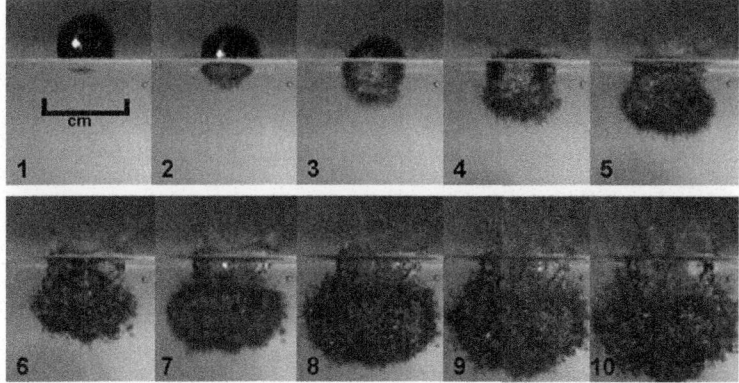

Fig. 4.23 High-speed photography of the fragmentation of a metal melt drop dripping into the liquid at 6,000 fps. Reprinted with the permission from Ref. [18] Copyright (2011) (ELSEVIER)

cooling process, the metal melt drop maintains a spherical shape, with only irregular needle-like protrusions appearing on the surface.

References

1. Okawa T, Kubo K, Kawai K et al. Experiments on splashing thresholds during single-drop impact onto a quiescent liquid film. Exp Thermal Fluid Sci 121
2. Liang G, Guo Y, Yang Y et al (2013) Special phenomena from a single liquid drop impact on wetted cylindrical surfaces. Exp Thermal Fluid Sci 51:18–27
3. Liang G, Guo Y, Yang Y et al (2014) Liquid sheet behaviors during a drop impact on wetted cylindrical surfaces. Int Commun Heat Mass Transfer 54:67–74
4. Almohammadi H, Amirfazli A (2019) Drop impact: viscosity and wettability effects on splashing. J Colloid Interf Sci 553:22–30
5. Burzynski DA, Roisman IV, Bansmer SE (2020) On the splashing of high-speed drops impacting a dry surface. J Fluid Mech 892
6. Zhuo Y-Y, Hussain S, Lin S-Y (2021) Effect of surface roughness on the collision dynamics of water drops on wood. Colloids Surf, A 612:125989
7. Painuly R, Kumar S, Anand V (2024) Coalescence of hydrolysed polyacrylamide and surfactant based drop on water–oil interface under an electric field. Colloids Surf A: Physicochem Eng Aspects 697
8. Chashechkin YD, Ilinykh AY (2023) Intrusive and impact modes of a falling drop coalescence with a target fluid at rest. Axioms 12(4)
9. Thoroddsen ST, Takehara K, Etoh TG (2005) The coalescence speed of a pendent and a sessile drop. J Fluid Mech 527:85–114
10. Gaitzsch F, Gäbler A, Kraume M (2011) Analysis of droplet expulsion in stagnant single water-in-oil-in-water double emulsion globules. Chem Eng Sci 66(20):4663–4669
11. Zhang Y, Zhang X, Zhang S et al. Research on dynamic process and drop splash of laser-induced cavitation bubble collapse within a drop. Appl Sci 13(13)
12. Zhang Y, Zhang X, Zhang S et al (2023) Research on eccentric cavitation bubble collapse dynamics within drops. Symmetry 15(7)

13. Avila SRG, Ohl CD (2016) Fragmentation of acoustically levitating droplets by laser-induced cavitation bubbles. J Fluid Mech 805:551–576
14. Thoroddsen ST, Etoh TG, Takehara K (2007) Microjetting from wave focusing on oscillating drops. Phys Fluids 19(5):052101
15. Van Hoeve W, Gekle S, Snoeijer JH et al (2010) Breakup of diminutive Rayleigh jets. Phys Fluids 22(12):122003
16. Zhao H, Nguyen D, Edgington-Mitchell D et al (2023) The largest diameter of falling drop in the up-gas flow. Int J Multiphase Flow 159
17. Zheng K, Gan Z (2024) Investigation in the bag breakup of water drop in airflow with different temperatures and densities. Phys Fluids 36(9)
18. Sa R, Takahashi M, Moriyama K (2011) Study on fragmentation behavior of liquid lead alloy drop in water. Prog Nucl Energy 53(7):895–901

Chapter 5
Visualization Research on Blunt Body Wake Dynamics

5.1 Introduction

This chapter investigates the wake dynamics formed downstream of a blunt body under various motion states. The wake evolution of the blunt bodies is analyzed. The typical blunt bodies include cylinders, spheres, and prisms. The motion of the blunt body is defined as its relative motion with the flow field. Thus, the static blunt bodies in uniform flow fields are classified as translational motion.

Mostly, the wake motion is very slow, and the shooting speed for the high-speed photography is also low (i.e. below 2000 fps). Only when it comes to cavitation vortex shedding, a higher shooting speed is needed. Furthermore, since the characteristics of the wake are not easily observable through conventional photography techniques, the particle tracking velocimetry (PTV), particle image velocimetry (PIV), and virtual dye visualization are widely applied for the wake tracing.

5.2 Cylinder

5.2.1 Translation

When a fluid flows past a cylinder, the flow characteristics of the fluid are altered, resulting in the formation of a lower velocity region downstream of the cylinder, known as the wake. The Reynolds number (Re) is introduced to describe the flow characteristics of the fluid.

Figure 5.1 illustrates the photographs of the cylinder wake at $Re = 300$, captured at 10,000 fps. In the figure, the classic Karman vortex street is observed. In subfigure (a), the wake behind the cylinder has reached a stable state. To further investigate the dynamic characteristics of the wake, a large number of particles are placed in the fluid. By observing the position of these particles within the wake, the detailed characteristics of the wake can be observed. In subfigure (b), the black area represents the large number of the particles injected into the fluid. At this point, the particles have not yet entered the wake region. As time progresses, in subfigure (c), the particles are entrained into the wake. In subfigure (d), most particles flow downstream with the wake, while a few particles are drawn into the vortex, which is more evident in subfigure (e). In subfigure (f), the particles circulate with the vortices, and some particles enter the vortices on the symmetric side. After several cycles, the particles finally detach from the vortices and continue to flow downstream.

Figure 5.2 illustrates the photographs of the cylinder wake at $L/D = 3$, captured at 800 fps. Here, the parameter L/D is defined to characterize the size ratio of the flexible film to the cylinder. L represents the length of the flexible film, and D represents the cylinder diameter. Two motion states of the flexible film in the cylinder wake are defined: the bent state and the unbent state. The figure utilizes virtual dye visualization technology to observe the structure of the cylinder wake. This technique allows a controlled number of the particles to be placed at any location and colored based

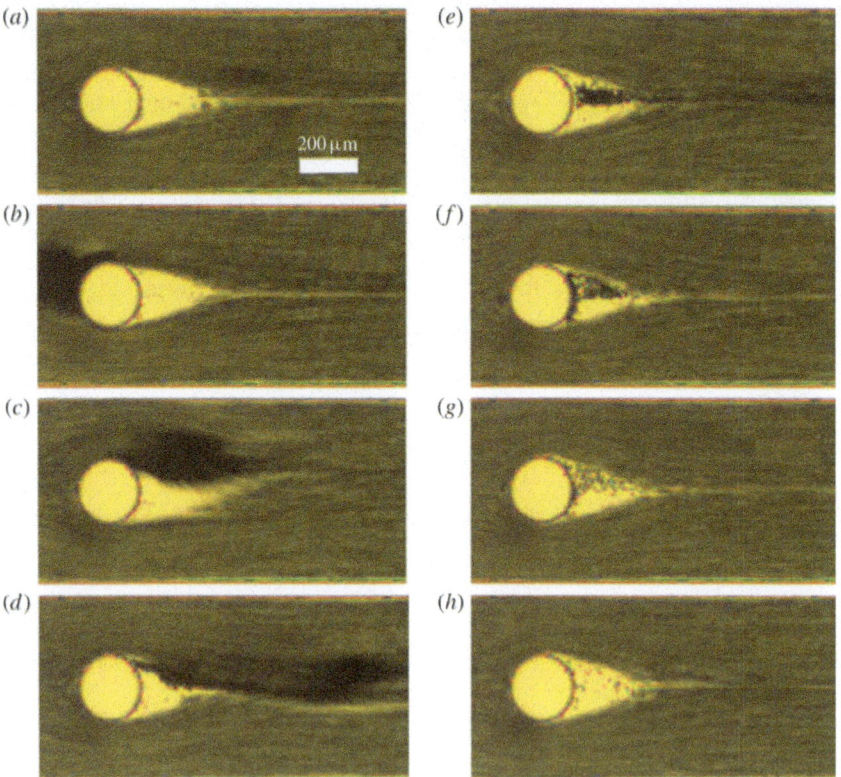

Fig. 5.1 High-speed photography of the cylinder wake at Re=300, captured at 10,000 fps. Reprinted with the permission from Ref. [1] Copyright (2014) (Cambridge University Press)

5.2 Cylinder

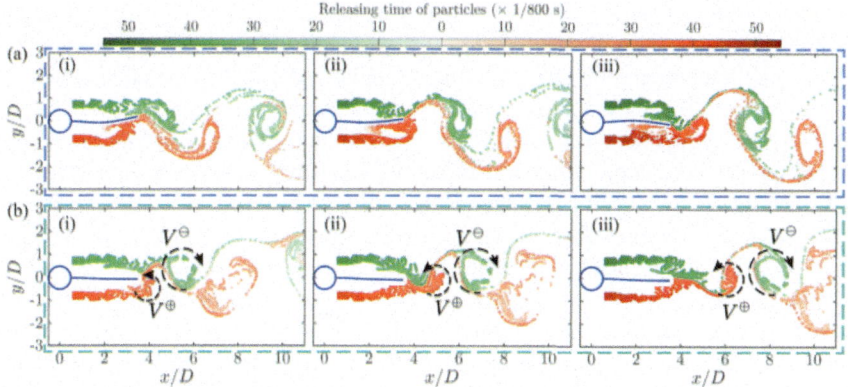

Fig. 5.2 High-speed photography of the cylinder wake at $L/D = 3$, captured at 800 fps. Reprinted with the permission from Ref. [2] Copyright (2024) (AIP Publishing)

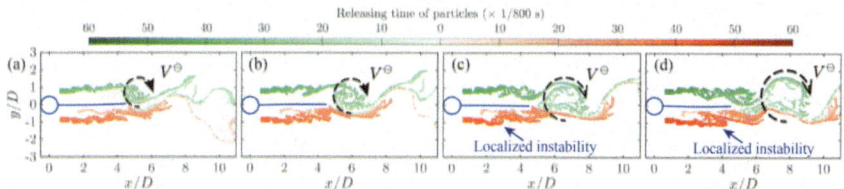

Fig. 5.3 High-speed photography of the cylinder wake at $L/D = 4$, captured at 800 fps. Reprinted with the permission from Ref. [2] Copyright (2024) (AIP Publishing)

on their placement position. In Fig. 5.2, the green particles are from the top the red particles are from the bottom. In subfigure (a), the flexible film is in a bent state, with the vortex on the downstream side of the cylinder. In subfigure (b), the flexible film is in an unbent state, with the vortex forming at the free end of the flexible film.

Figure 5.3 illustrates the photographs of the cylinder wake at $L/D = 4$, captured at 800 fps. Under this condition, the flexible film maintains an unbent state, with the vortices forming behind the free end of the flexible film. Over time, these vortices move downstream and the affected area expands. Compared to the case of $L/D = 3$, the wake behind the cylinder is relatively more stable.

5.2.2 Rotation

When the cylinder rotates, its wake exhibits distinct flow patterns. The dimensionless forced rotation frequency is defined as $f = f_f / f_0$, where f_0 is the shedding frequency of the cylinder wake vortices without forcing, and f_f is the forced rotation frequency of the cylinder. Additionally, A is defined as the dimensionless amplitude of the cylinder forced oscillation, calculated by the formula $A = D\theta_0 \pi f_f / U_0$. Here, D is the cylinder diameter, U_0 is the freestream velocity, and θ_0 is the initial angle. Figure 5.4 illustrates the photographs of the rotating cylinder wake at $A = 2$ with different f, captured at 250 fps. Laser-induced fluorescence technology is employed to track the structure of the cylinder wake. As shown in subfigure (a) ($f = 0.5$), the forcing frequency is lower than the natural shedding frequency. Vortices are observed on the downstream side of the cylinder. After the vortices shed, a series of vortices with alternating rotation directions is formed. In subfigure (b) ($f = 1$), the forcing frequency matches the natural shedding frequency, which is the resonance condition. Similar to the vortex motion in subfigure (a), the spacing between the two vortices of the wake is shortened. In subfigure (c) ($f = 2$), a boundary line appears in the cylinder wake, starting from the center of the cylinder and extending horizontally downstream. In subfigure (d) ($f = 3$), the vortices are distributed above and below the boundary line, and the shape of the vortices becomes less distinct. In subfigure (e) ($f = 4$), the boundary line of the wake shortens, and more vortices are formed behind the wake. In subfigure (f), $f = 5$, the boundary line further shortens.

Figure 5.5 illustrates the photographs of the rotating cylinder at $f = 5$ under different forced amplitudes, captured at 250 fps. As shown in subfigure (a) ($A = 1$), there is a section of demarcation on the downstream side of the cylinder, with incomplete vortices forming on either side. Behind the demarcation region, distinct vortices are formed and alternate with adjacent vortices rotating in the opposite direction. In subfigure (b) ($A = 2$), the wake structure is similar to that in subfigure (a), while the vortices become more compact. In subfigure (c) ($A = 2$), the demarcation region extends downstream, and no complete vortex is formed behind the demarcation region. In subfigure (d) ($A = 4$), the demarcation region further extends. In subfigure (e) ($A = 7$), the demarcation region shortens, and the vortices appear downstream again. In subfigure (f) ($A = 9$), the demarcation region is not significantly varied. And, different from subfigure (e), the vortices on the upper and lower sides of the demarcation region converge at the end of it.

5.3 Sphere

5.3.1 Rolling

Spheres are also one of the typical bluff bodies. The movement pattern of the sphere and its interaction with the incoming flow greatly affect the formation and evolution of the wake. Figure 5.6 illustrates the photographs of the sphere wake at Re = 820, captured at 2,000 fps. The sphere undergoes rolling motion near a flat wall with a slow speed. It can be observed that the vortices shed downward from the top of the sphere, and interact with the vortices shed from the bottom. Thus, the wake presents a horseshoe shape. These vortices continuously reorganize as they propagate downstream, with their rotation axes constantly changing. Notably, in subfigure (f),

5.3 Sphere

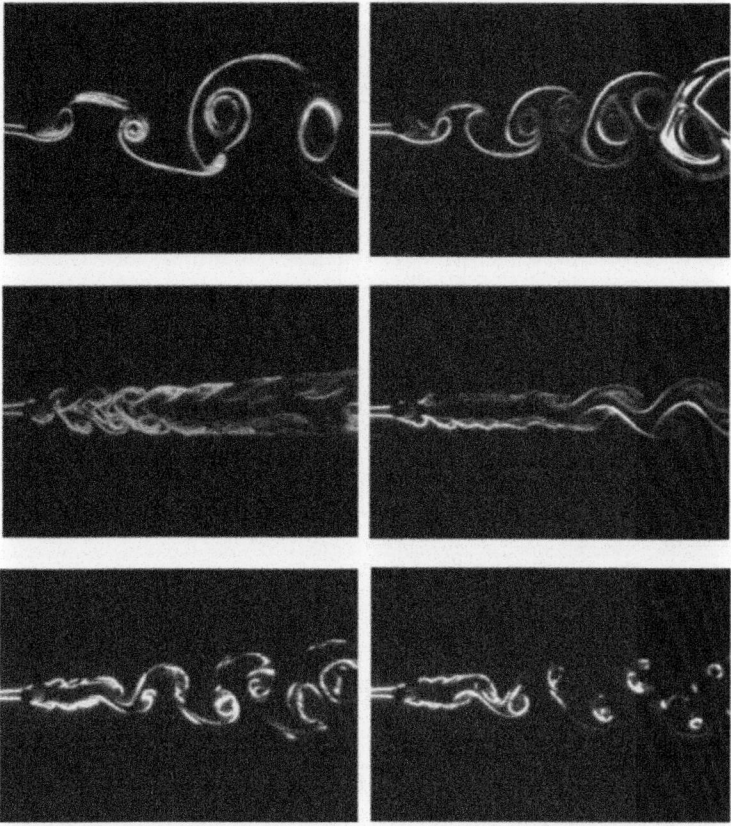

Fig. 5.4 High-speed photography of the rotating cylinder wake with different forcing frequency, captured at 250 fps. Reprinted with the permission from Ref. [3] Copyright (2006) (Cambridge University Press)

the width of the sphere wake is observed to gradually increase as it moves further downstream, with the vortices becoming more diffuse.

Figure 5.7 illustrates the photographs of the sphere wake at Re = 2020, captured at 2,000 fps. As shown in the figure, vortices shed from the top and merge with those shed from the bottom, forming larger vortex structures. However, due to the strong interaction between the vortex rings and the wall, the vortex rings are disrupted, forming smaller vortex structures. At the end of the wake, these smaller vortices further mix and break down, presenting turbulent characteristics. In subfigure (f), the wake spreads downstream along the flow direction, with the increasing width. Compared to Fig. 5.6, at a higher Re, the sphere wake is more complex and exhibits richer flow characteristics.

In addition, the wake characteristics of a rolling sphere are significantly influenced by the incoming flow velocity and the rotation frequency. The dimensionless rolling speed is defined as $\Omega = U_r/U$, where U_r is the surface speed of the sphere and U

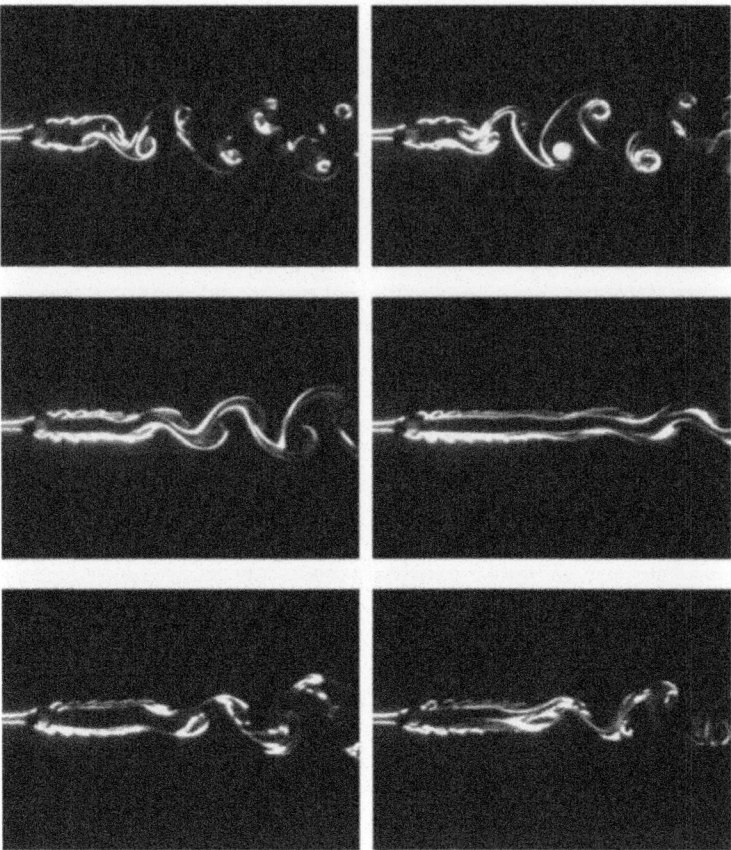

Fig. 5.5 High-speed photography of the rotating cylinder wake with different forcing amplitude, captured at 250 fps. Reprinted with the permission from Ref. [3] Copyright (2006) (Cambridge University Press)

is the incoming flow velocity. Figure 5.8 shows the high-speed photography of the sphere wake at $\Omega = 0.375$, captured at 2,000 fps. In the figure, the structure of the sphere wake is presented through the dye visualisation technique. At Re = 115, the wake exhibits a wavy flow structure that extends downstream. As Re increases to 150, vortices are observed to shed from both the top and bottom of the sphere, forming the horseshoe shape with two concave surfaces. As Re continues to increase, the shedding frequency of the horseshoe vortices rises. At Re = 250, the flow separation occurs on both sides of the sphere, resulting in alternating horseshoe vortices. This alternating shedding pattern becomes more pronounced with further increases in Re (i.e., Re = 320). When Re reaches 450, the horseshoe vortices become less prominent, and the flow structure begins to break down, which is even more evident at Re = 550.

Figure 5.9 illustrates the photographs of the sphere wake at $\Omega = 0.5$, captured at 2,000 fps. As shown in the figure, at Re = 200, vortices shed from the top of the

5.3 Sphere

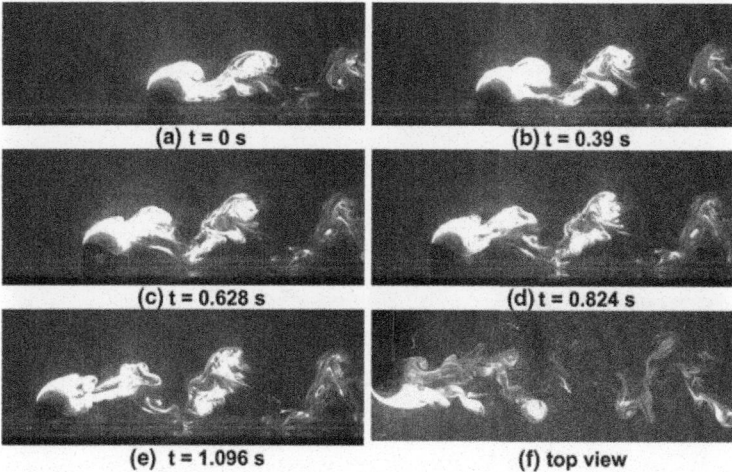

Fig. 5.6 High-speed photography of the sphere wake at Re = 820, captured at 2,000 fps. Reprinted with the permission from Ref. [4] Copyright (2021) (Springer Nature)

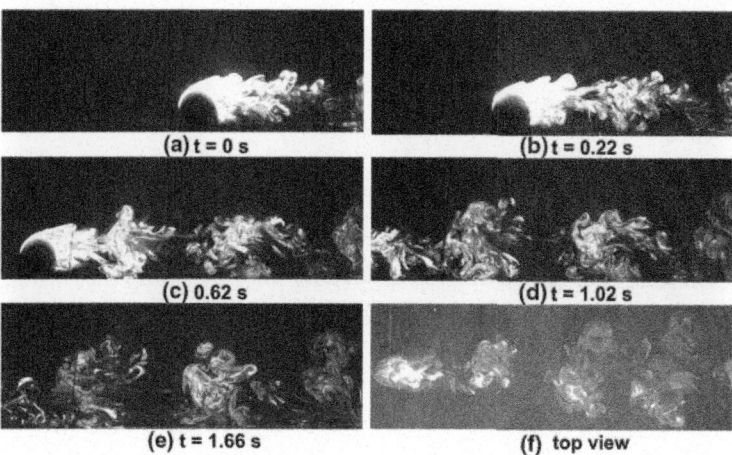

Fig. 5.7 High-speed photography of the sphere wake at Re = 2020, captured at 2,000 fps. Reprinted with the permission from Ref. [4] Copyright (2021) (Springer Nature)

sphere, forming a unilateral and elongated horseshoe vortex. And, the bottom of the sphere, influenced by the wall, does not produce any vortices. As Re increases to 250, vortices begin to shed alternately from both sides of the sphere. At Re = 300, these alternating vortices from both sides form bilateral horseshoe vortices. As Re continues to increase, the flow structure starts to be disrupted, and the frequency of the vortex shedding also increases. At Re = 550, the wake behind the sphere exhibits turbulent characteristics, becoming extremely unstable. The large vortices

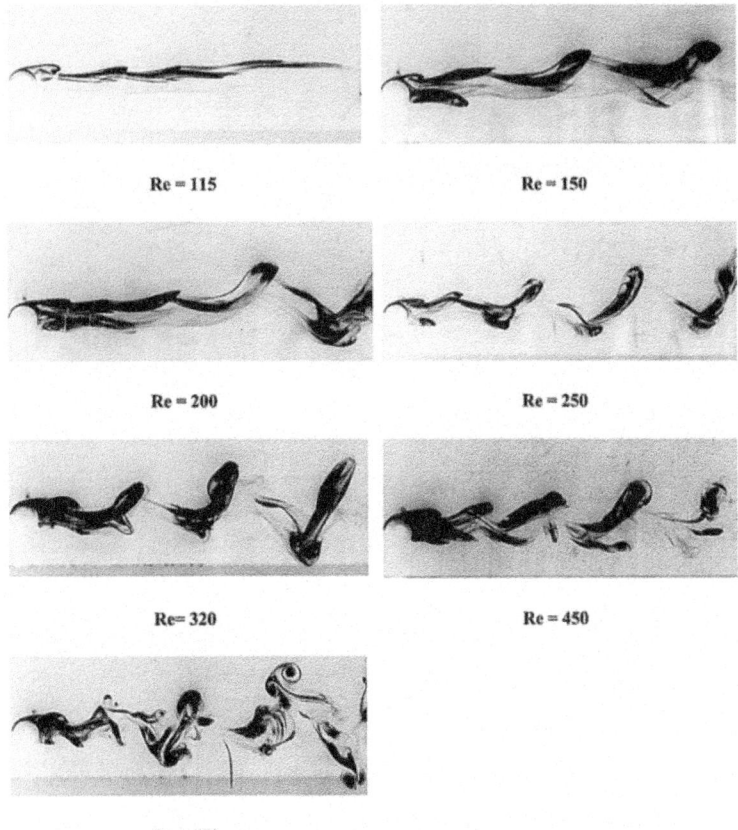

Fig. 5.8 High-speed photography of sphere wake at $\Omega = 0.375$, captured at 2,000 fps. Reprinted with the permission from Ref. [5] Copyright (2021) (Springer Nature)

break down into smaller ones, then merge and link with each other, presenting an irregular flow pattern.

5.4 Prism

5.4.1 Translation

Due to its unique structure, the triangular prism is prone to cavitation downstream. Cavitation number (σ) is employed to describe the cavitation state. Figure 5.10 illustrates the photographs of the triangular prism wake at $\sigma = 2.7$, captured at 5,000 fps. The cavitation regions appear on the downstream side of the triangular

5.4 Prism

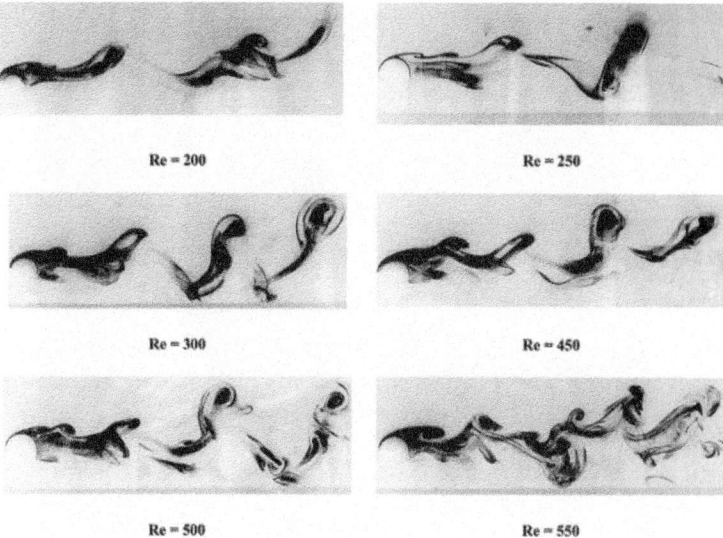

Fig. 5.9 High-speed photography of sphere wake at $\Omega = 0.5$, captured at 2,000 fps. Reprinted with the permission from Ref. [5] Copyright (2021) (Springer Nature)

prism, and then detach to form vortices that propagate downstream. And, the vortices spread into cloud-like structure, increasing the width of the wake. Simultaneously, vortices are observed to alternately distribute from the triangular prism, with adjacent vortices rotating in opposite directions (i.e., subfigure (b). In subfigure (a), the detached vortices exhibit a wavy pattern as they propagate downstream, and eventually transform into cloud-like vortices.

Figure 5.11 illustrates the photographs of the triangular prism wake at $\sigma = 1.8$, captured at 5,000 fps. In subfigure (a), a large cavitation region is observed on the downstream side of the triangular prism, and then propagates downstream in a ribbon-like structure. These ribbon-like vortices become thinner as they propagate. In subfigure (b), the cavitation region takes on the shape of the triangular prism. During propagation process, part of the vortices separates from the main vortex region, causing the wake wider. Compared to Fig. 5.10, the distribution of the vortices does not show a clear upper and lower spacing. Notably, adjacent vortices are connected by multiple cavitation tubes.

The quadrangular prism exhibits a particularly significant obstructive effect in the fluids. The wake characteristics of the quadrangular prism vary greatly at different heights from the flat wall. To describe the relative position of the quadrangular prism to a flat wall, the dimensionless height is defined as $z^* = H/d$. Here, H is the observation height and d is the edge length of the prism top surface. Figure 5.12 shows the high-speed photography of the quadrangular prism wake at different heights, captured at 20,000 fps. The quadrangular prism in the figure has a flexible film attached to the top. The dimensionless length is defined as $l^* = l/d$ to characterize

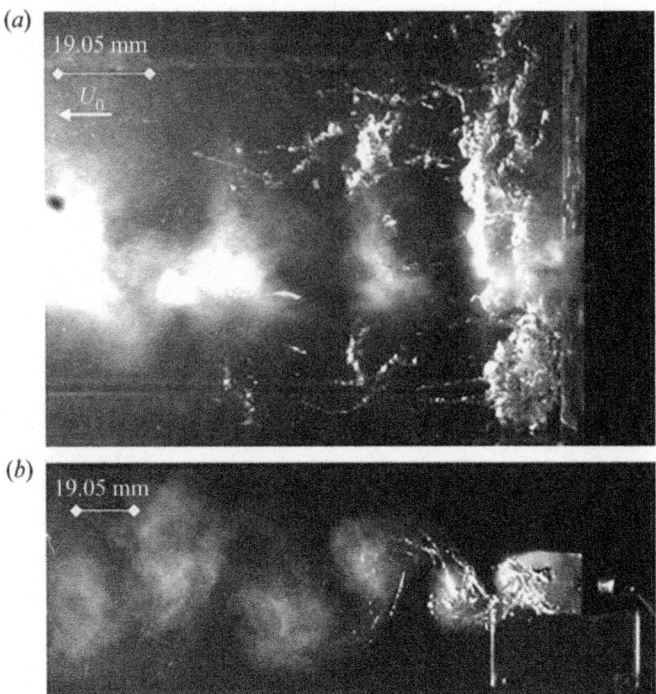

Fig. 5.10 High-speed photography of triangular prism wake at σ = 2.7, captured at 5,000 fps. Reprinted with the permission from Ref. [6] Open access under a CC BY 4.0 license, https://creativecommons.org/licenses/by/4.0/

the length of the film. Here, l is the length of the flexible film. Based on the smoke-wire flow visualization technology, the structure of the quadrangular prism wake is captured. As shown in the figure, at a small z^*, a cavitation region appears on the downstream side, and the fluid velocity is relatively low. And, vortices form on both sides of the cavitation region, and then merge. At $z^* = 4$, this phenomenon is the most apparent. As z^* increases to 4.5, the cavitation region on the downstream side disappears, and vortices on the upper and lower sides separate and fill the region. As z^* further increases, the vortices gradually disappear, and the flow transitions to the laminar state. When comparing the situations with and without the flexible film, it is found that the presence of the flexible film significantly disrupts the structure of the wake, which is particularly noticeable when z^* is small.

As the number of the prism edge increases, its obstructive effect on the incoming flow also decreases. And, its wake structure becomes more similar to that of a cylinder. Figure 5.13 shows the high-speed photography of the polygonal prism wake, captured at 200 fps. Particle image velocimetry (PIV) is applied to track the wake structure, and appropriate orthogonal decomposition techniques are employed to extract information about the vortices. Subfigure (a) illustrates the wake of a

5.4 Prism

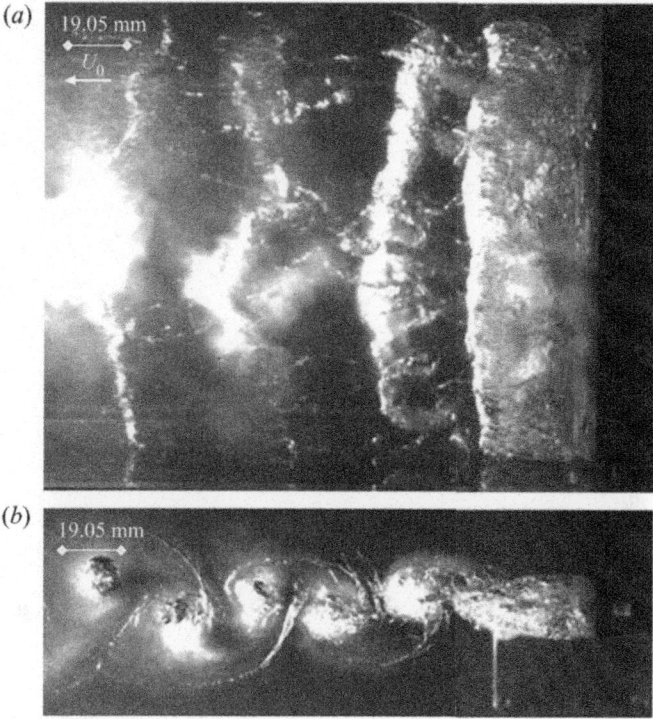

Fig. 5.11 High-speed photography of triangular prism wake at $\sigma = 1.8$, captured at 5,000 fps. Reprinted with the permission from Ref. [6] Open access under a CC BY 4.0 license, https://creativecommons.org/licenses/by/4.0/

cylinder, where vortices are shed alternately. Subfigure (b) presents the polygonal prism wake, which also exhibits alternating vortex shedding. Subfigure (c) shows the polygonal prism wake with surface oscillation, which has similar characteristics to the polygonal prism wake, but with more rapid vortex shedding. Although both the polygonal prism wake and the cylinder wake demonstrate the phenomenon of alternating vortex shedding, the morphology of the shear layers differs. In subfigure (a), the upstream and downstream shear layers gradually converge in the downstream direction. In subfigure (b), the shear layers remain approximately horizontal. In subfigure (c), the shear layers are very short, and vortex shedding occurs immediately after the fluid passes over the prism.

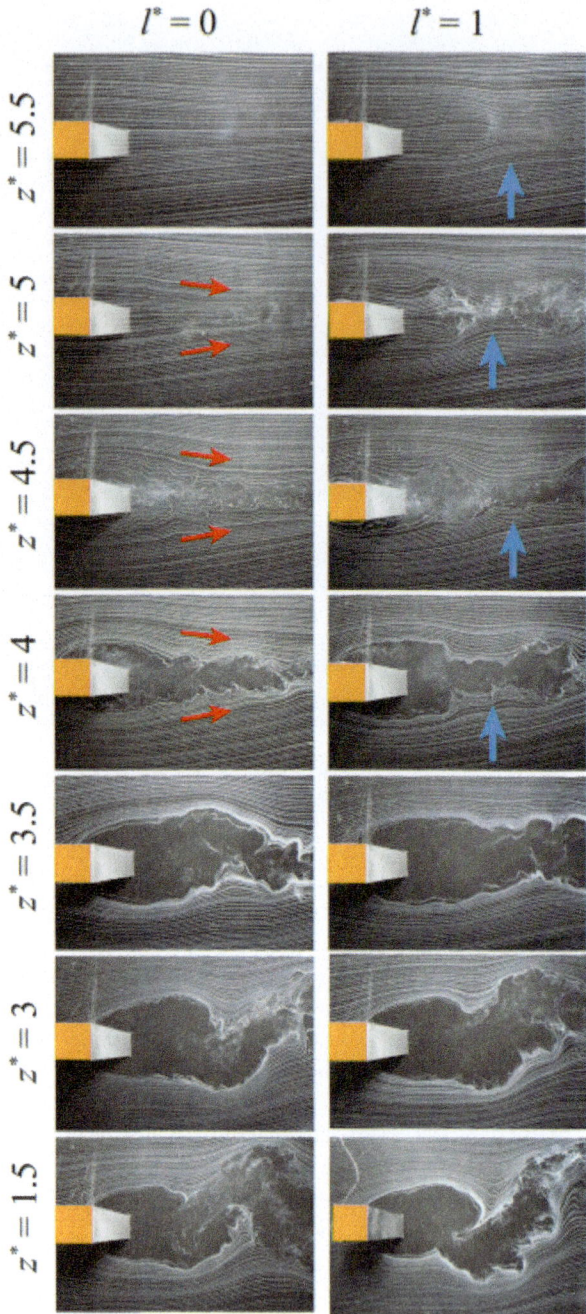

Fig. 5.12 High-speed photography of the quadrangular prism wake with different heights, captured at 20,000 fps. Reprinted with the permission from Ref. [7] Copyright (2023) (AIP Publishing)

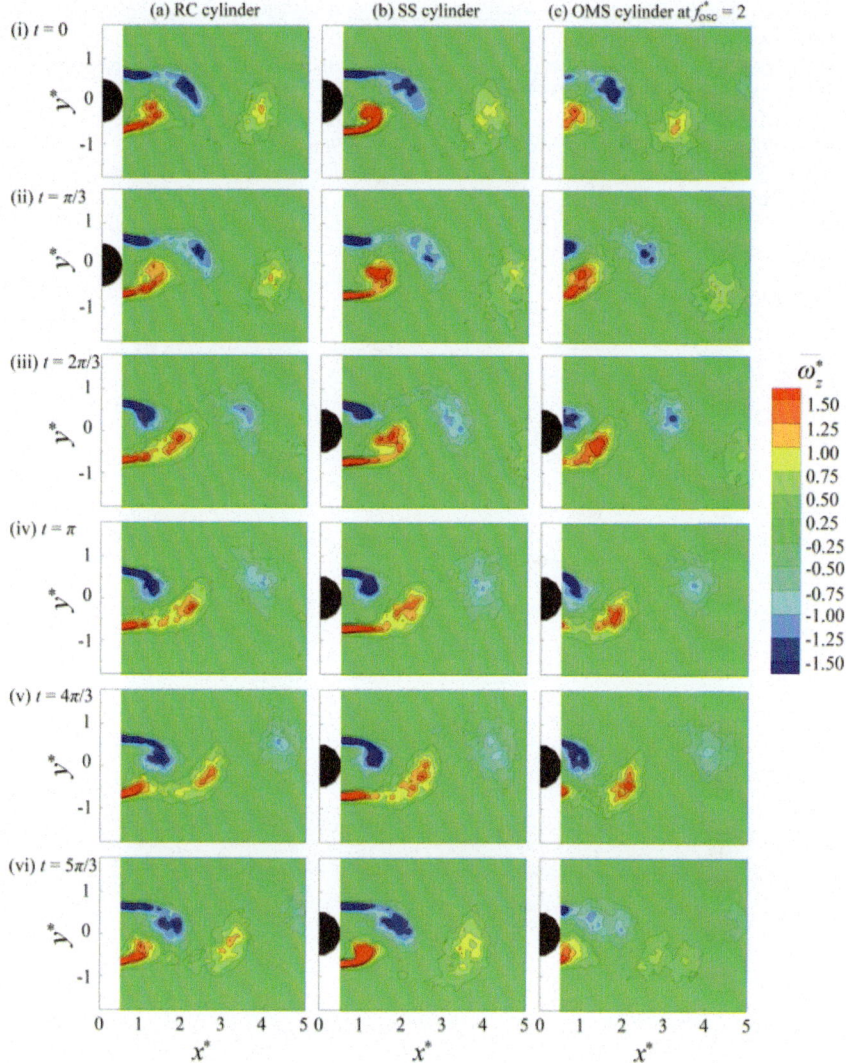

Fig. 5.13 High-speed photography of the polygonal wake at 200 fps. Reprinted with the permission from Ref. [8] Copyright (2024) (AIP Publishing)

References

1. Haddadi H, Shojaei-Zadeh S, Connington K et al (2014) Suspension flow past a cylinder: particle interactions with recirculating wakes. J Fluid Mech 760
2. Duan F, Wang J-J (2024) Mode transition of a film fluttering in a circular cylinder wake. Phys Fluids 36(5)

3. Thiria B, Goujon-Durand S, Wesfreid JE (2006) The wake of a cylinder performing rotary oscillations. J Fluid Mech 560
4. Chandel A, Das SP (2021) Effect of wall proximity on the wake of a rotating and translating sphere. Acta Mech 232(12):4833–4846
5. Chandel A, Das SP (2021) Wake of transversely rotating and translating sphere in quiescent water at low Reynolds number. Acta Mech 232(3):949–966
6. Wu J, Deijlen L, Bhatt A et al (2021) Cavitation dynamics and vortex shedding in the wake of a bluff body. J Fluid Mech 917
7. Zhao C, Wang H, Liu Z et al (2023) Near-wake structures of a finite square cylinder with a flapping film at its free end. Phys Fluids 35(9)
8. Zeng L, New TH, Tang H (2024) Control of cylinder wake using oscillatory morphing surface. Phys Fluids 36(5)

Chapter 6
Conclusion

This book comprehensively reviews the applications of the high-speed photography in fluid mechanics, showcasing various transient phenomena in fluid motion. Additionally, visualization research on the bubble, drop, and wake dynamics are presented, aided by techniques such as schlieren photography and particle image velocimetry. Conclusions are given as follows.

Firstly, the development history of the high-speed photography technology is introduced, as well as its application scenarios in experimental fluid mechanics. And, the future development of the high-speed photography technology lies in achieving faster speeds, clearer images, and the ability to capture microscopic objects, such as the high-speed digital holography and nanoscale high-speed observation techniques.

Secondly, the bubble dynamics in various scenarios are introduced. The morphology evolution, collapse jet, and shock wave near various boundaries are analyzed. In the visualization research on the bubble dynamics, the speeds of the high-speed photography are mostly below 300,000 fps. And, the multi perspective shooting, and schlieren techniques are applied to explore the mechanisms of the bubble dynamics.

Thirdly, the dynamic behaviors of the drop undergoing different motions are introduced. The impact and coalescence motions of the drops generally do not require very high shooting speeds (i.e., lower than 10,000 fps). When it comes to the microjet phenomena, due to the rapid generation and movement, higher shooting speeds (i.e., higher than 100,000 fps) are required.

Finally, the wake dynamics formed downstream of the blunt bodies under various motion states are introduced. The development and evolution of the wake are analyzed during the translational and rotational motions of the blunt bodies. Mostly, the wake motion is very slow, and the shooting speed for the high-speed photography is also low (i.e. below 2000 fps). Only when it comes to the cavitation vortex shedding, a higher shooting speed is needed. Technologies such as particle tracking velocimetry, particle image velocimetry, and virtual dye visualization are widely applied for the wake tracing.

Index

B
Blunt body, 2, 4, 34, 35, 57

C
Cavitation bubble, 1, 26, 31
Cavitation flow, 2–4, 11, 29, 30, 32, 33
Coalescence, 2–4, 37, 44, 46–48, 71
Collapse, 1–3, 8, 15–28, 31, 32, 48–50, 71
Cylinder, 1, 4, 24, 25, 40, 41, 57–62, 66, 67

D
Drop, 1–4, 8, 9, 24, 25, 29, 37–55, 71

F
Fragmentation, 1–4, 37, 52, 53, 55

H
High-speed photography, 1–4, 7–9, 11, 15–35, 38–55, 57–59, 61–69, 71

J
Jet, 1–4, 8, 15, 18–25, 27, 31, 34, 48, 50, 51, 71

M
Motion, 1–4, 8–10, 15, 16, 22, 26–28, 37, 38, 57, 58, 60, 71

N
Narrow gap, 4, 22–25, 28, 29

P
Particle, 2, 3, 8, 9, 15, 22, 26–28, 57, 66, 71
Pinch-off, 51, 52
Prism, 4, 8, 57, 64–68

S
Shock wave, 2, 3, 10, 15, 20–22, 27, 71
Sphere, 4, 8, 57, 60–65
Splash, 3, 8, 42, 48, 50

V
Vortex, 1, 4, 23, 31–33, 45, 57–61, 63, 65, 67, 71

W
Wake, 2, 4, 9, 57–69, 71

The manufacturer's authorised representative in the EU is Springer Nature Customer Service Centre GmbH, Europaplatz 3, 69115 Heidelberg, Germany. If you have any concerns regarding our products, please contact ProductSafety@springernature.com

Printed and bound by CPI Group (UK) Ltd, Croydon, CR0 4YY
26/03/2026
02078940-0008